Investigating Probability and Statistics Using the TI-82 Graphics Calculator

Student Edition

Graham A. Jones
Beverly S. Rich
Carol A. Thornton
Roger Day

Dale Seymour Publications®

We wish to thank Michelle Bell, Elaine Caspers, Dana Jensen, Marion Jones, Sheryl Kidd, and Marilyn Parmantie, all of whom contributed to the books in a variety of ways.

Managing Editor: Cathy Anderson
Project Editor: Katarina Stenstedt
Production: Leanne Collins
Design Manager: Jeff Kelly
Text/Cover Design: London Road

This book is published by Dale Seymour Publications®, an imprint of Addison Wesley Longman, Inc.

Printed in the United States of America.

ISBN 0-201-49353-5

5 6 7 8 9 10-ML-00

Contents

To the Student vii

Module 1: Analyzing Data

Looking Ahead 2

Starting

1-1 What's in a Name? 3
1-2 Names in a Histogram 4
1-3 What's the Mean for the "Name" Data? 5
1-4 What's the Median for the "Name" Data? 6
1-5 Quartering the "Name" Data 7
1-6 What's the Mode for the "Name" Data? 8
1-7 Measuring the Spread of the "Name" Data 9
1-8 Who's Far Out? 10

Developing

1-1 Games Won 11
1-2 Runs Scored 12
1-3 Truncate or Round? 13
1-4 Comparing Runs 15
1-5 Boxing Games Won 16
1-6 Comparing Games Won 18
1-7 Runs Scored and Games Won 19
1-8 Names and Shoe Lengths 20

Extending

1-1 Roll a Dozen 21
1-2 Who Was the Top Female Tennis Player? 22
1-3 The Top Ten Roadsters 23
1-4 The Speeds of Animals 24
1-5 Endangered Animals 25

Looking Back 26

Module 2: Probability, Simulations, and Random Numbers

Looking Ahead 28

Starting

2-1 Flip the Coin 29
2-2 Model the Coin Flip 31
2-3 Modeling Two Flips 32
2-4 Steffi's Serve 33
2-5 Flip the Cup 34
2-6 Chances for the Name 35
2-7 Modeling the Chances for the Name 36

2-8 Two Broadway Favorites ... 37
2-9 Three in a Row ... 38
Developing
2-1 Three Hits ... 39
2-2 The Top Ten ... 40
2-3 On Top for Four Weeks ... 41
2-4 Basketball Playoffs ... 42
2-5 Grand-Slam Match ... 43
2-6 Chasing the Circle ... 44
2-7 Estimating π ... 46
Extending
2-1 Bazuka Bats ... 47
2-2 Rock-Star Cards ... 48
2-3 Raining in Surfers' Paradise ... 49
Looking Back ... 50

Module 3: Counting

Looking Ahead ... 52
Starting
3-1 Naming the Dog ... 53
3-2 How Many Faces? ... 54
3-3 Birth Month in a Box ... 55
3-4 Three the Same ... 56
3-5 First Names in a Box ... 57
3-6 How Many Sandwiches? ... 58
3-7 How Many Kinds of Pizza? ... 59
Developing
3-1 Batting Order ... 60
3-2 Pitcher Bats Last ... 61
3-3 Raffle Tickets ... 62
3-4 Calculating the Number of Raffle Tickets ... 63
3-5 "Special" Raffle Tickets ... 64
3-6 "Extra-Special" Raffle Tickets ... 65
3-7 Mini-Lotto ... 66
3-8 Calculating the Mini-Lotto Pairs ... 67
3-9 Pick Three ... 68
3-10 Pick Three Again ... 69
Extending
3-1 Many the Same ... 70
3-2 Two in the Same Month ... 71
3-3 The State Lottery ... 72
3-4 Winning Hands ... 73
Looking Back ... 74

Module 4: Modeling and Predicting

Looking Ahead ... 76
Starting
4-1 Shoe and Arm Lengths ... 77
4-2 Shoe and Name Lengths ... 79
4-3 The Missing Movie ... 80

Developing
4-1 Shoe Lengths and Consonants 81
4-2 The Line of Best Fit 82
4-3 Equation for the Line of Best Fit 83
4-4 Graphing on the Scatter Plot 84
4-5 Finding the Correlation 85
4-6 Money Made and Movie Ranking 86
4-7 Money Made and Year Released 87
4-8 Forecasting the Missing Movie 88
Extending
4-1 Shoe Lengths and Arm Lengths 89
4-2 Testing the Radius and Area Relationship 90
4-3 Transforming the Radius and Area Data 91
4-4 Fitting Radius to Area 92
4-5 Testing the Radius and Volume Relationship 93
4-6 Crude-Oil Production 94
4-7 Transforming the Crude-Oil Production Data 95
4-8 Fitting Year and Crude-Oil Production 96
4-9 An Alternative Transformation 97
4-10 Another Transformation 98
4-11 Non–English-Speaking Americans 99
4-12 How the United States Has Grown 100
Looking Back 101

Appendix A: Blackline Masters

Blackline Master 1: Circle Mats 104
Blackline Master 2: Names and Birth Months 105
Blackline Master 3: More Names and Birth Months 106
Blackline Master 4: Quarter-Inch Graph Paper 107
Blackline Master 5: Circles 108

Appendix B: Calculator Helps

Calculator Help 1: Entering Data 110
Calculator Help 2: Range; Histogram 111
Calculator Help 3: Mean 113
Calculator Help 4: Generating a Scatter Plot 114
Calculator Help 5: Using the Random-Number Generator 115
Calculator Help 6: Using the Line Function; Cursor Movement in the Line Function 116
Calculator Help 7: Finding the *y*-Intercept, the Slope of the Line of Best Fit, and the Correlation Coefficient (*r*); Using the Equation to Draw the Line of Best Fit 117
Calculator Help 8: Using the Best-Fit Equation to Predict a *y* Value for a Given *x* 118
Calculator Help 9: Entering Transformed Data 119

Appendix C: Calculator Programs

Introduction 121
Calculator Program 1: Stem-and-Leaf Plot 125
Calculator Program 2: Ten 127
Calculator Program 3: Hundred 128

Appendix D: Data Sets

Data Set 1: First Names of Students ... 130
Data Set 2: American League/National League Statistics (1993) ... 131
Data Set 3: First Names of Students and Their Shoe Lengths ... 132
Data Set 4: The Top Women's Tennis Money Winners (1993) ... 133
Data Set 5: The Ten Top-Selling Passenger Cars in the United States by Calendar Year (1989–1991) ... 134
Data Set 6: The Speeds of Animals ... 135
Data Set 7: United States List of Endangered Species ... 136
Data Set 8: Broadway Favorites ... 137
Data Set 9: Pop Albums ... 138
Data Set 10: Top Money-Making Movies of All Time ... 139
Data Set 11: Annual World Crude-Oil Production (1880–1992) ... 140
Data Set 12: Number of Non–English-Speaking Americans ... 141
Data Set 13: Population of the United States from 1790–1990 ... 142

Appendix E: Word Bank ... 143

References ... 148

To the Student

This book offers a series of exploratory activities on probability and statistics using the TI-82 graphics calculator. It is intended to be used as a supplement to your regular math program and will give you the opportunity to investigate important applications and concepts that are now possible with graphics technology.

While most students will generally use this book as part of their ongoing classroom math program, it is possible to do the activities independently or in collaborative groups. Look at the reference list at the end of this book if you wish to pursue topics in greater depth or to enhance your own knowledge in a particular area.

USING THE BOOK

There are four modules in this book: Analyzing Data; Probability, Simulations, and Random Numbers; Counting; and Modeling and Predicting. Each module begins with a *Looking Ahead* section that provides an overview of the kind of activities you will explore during the module. A *Looking Back* section concludes each module and enables you to assess your personal understanding of the key concepts in the module and to share your thinking with others.

Each module has three levels of activity—Starting, Developing, and Extending—to accommodate different background experiences. Each of these levels includes a series of activities and appropriate calculator instructions.

If you are a beginner you will find it more appropriate to begin with the Starting activities. If you have had substantial experience you may find that you can move on to the Developing activities or even the Extending activities. In any case, wherever you start you may need to check back on some of the technology techniques that are introduced at various stages of the book.

SPECIAL FEATURES

You will find the following features helpful in using the book:

- **The Need List.** At the bottom of each activity page, all materials needed for the activity are listed. It is important to check this list and gather all the materials before beginning an activity.

- **Blackline Masters.** In some activities you will need special mats, cards, or graph paper. These are provided in Appendix A.

- **Calculator Helps.** These pages explain how to carry out particular procedures or techniques on your graphics calculator. They can be found in Appendix B. The pages are referenced in the activities where the procedures are needed.

- **Calculator Programs.** Some statistical processes and visual presentations will require more detailed programming. Instructions on how to insert and execute these programs are contained in Appendix C. These are referenced, when needed, in the activities.

- **Data Sets.** Even though data sets are provided for any activities that need them, you are encouraged to gather your own data on a number of occasions. Data sets are found in Appendix D.

- **Word Bank.** A glossary of terms commonly used in probability and statistics can be found in Appendix E. Reference is made to the "Word Bank" in relevant activities.

In this technological age data and chance situations are continually impacting our lives. Perhaps your experiences in this book using a graphics calculator will enable you to enjoy working with probability and statistics and to gain fresh insights into their applications.

Module 1

Analyzing Data

Looking Ahead

In this module you will explore real-world situations that involve data sets containing one or more variables. The major emphasis is on presenting the set of data in a way that makes it easy to interpret and to compare to similar sets of data. Several kinds of data presentation, such as stem-and-leaf plots and box plots, are considered.

You will use the graphics calculator to construct various visual presentations of data, and you will examine under what conditions these particular kinds of presentations are applicable. Measures of central tendency and measures of dispersion are used to describe, interpret, and analyze the data that have been drawn from a variety of social settings.

What's in a Name?

How many consonants are in first names?

Gathering and Entering Data

Organize

- Use "Data Set 1: First Names of Students" (page 130) or use the first names of students in your class.
- Enter the numbers of consonants in each first name into the TI-82 as the L_1 values. (See "Calculator Help 1: Entering Data" page 110.)

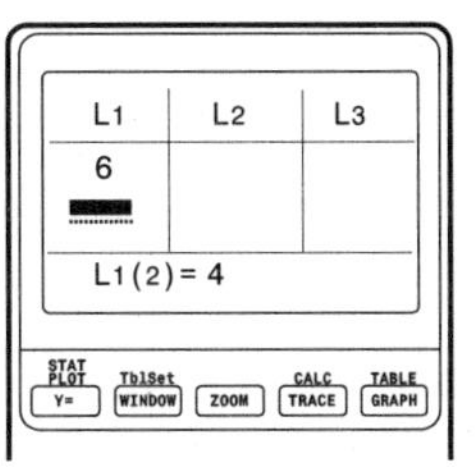

- Sort the data by pressing the following key sequence:
  STAT **2** (to select **2:SortA(** from the STAT EDIT menu)
  2nd [L1]) (to specify the list as L1)
  ENTER (to complete the process)

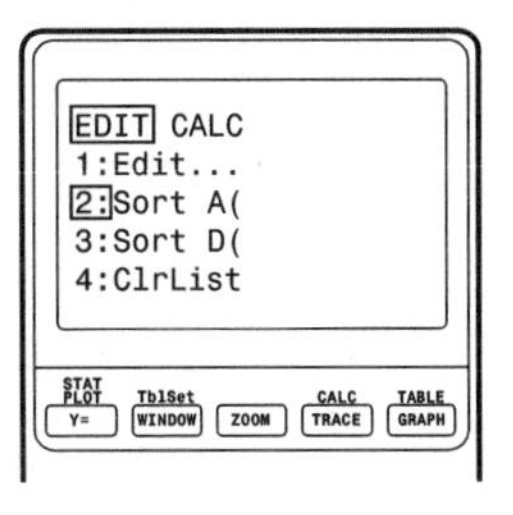

- Reexamine the data. Use the following key presses:
  STAT (to display STAT EDIT menu)
  ENTER (to select **1:Edit**)

Communicate

- How were the data in the display changed?

Reflect

- What does the set of data tell you?
- How might the set of data be displayed to make it easier to analyze? Would a bar graph help? Explain.

Need "Data Set 1: First Names of Students"

Names in a Histogram

For the "What's in a Name?" data, how might a histogram look?

Making a Histogram

Organize

- Use the WINDOW key to set the range for the data from Starting 1-1. (See "Calculator Help 2: Range; Histogram," page 111.)
- Make the histogram:
  [2nd] [STAT PLOT] (to display the STAT PLOTS menu)
  1 (to select **1:Plot1**)
  [<] [ENTER] (to turn plot on)
  [v] [>] [>] [>] [ENTER] (to highlight and select the histogram)
  [GRAPH] (to draw the histogram)

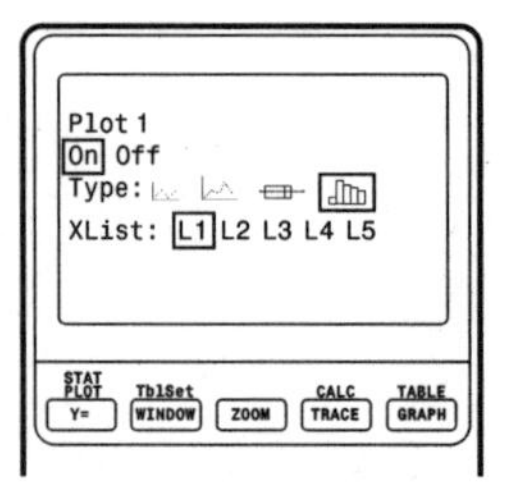

Communicate

- What does the histogram tell you about the number of consonants in people's names?
- Are there any clusters? gaps? outliers?

Reflect

- What would the histogram look like if you entered data on everyone in your school?
- Would the histogram portray the same information if you changed the width of the bar? (See "Calculator Help 2: Range; Histogram," page 111.)

Need Sorted data from Starting 1-1 "What's in a Name?"

What's the Mean for the "Name" Data?

What's the mean number of consonants in the first names?

Finding the Mean

Organize

- Display the sorted data from Starting 1-1 on the screen:
  STAT (to display STAT EDIT menu)
  ENTER (to select **1:Edit**)
- Use v to view the data (x values).
- Calculate the mean mentally. Check it using your calculator. (See "Calculator Help 3: Mean," page 113.)

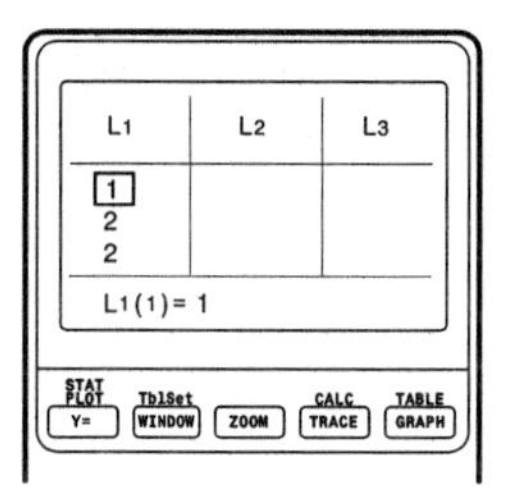

Communicate

- Was the mean a whole number?
- How would you interpret this mean if it were not?

Reflect

- If the student with the smallest number of consonants left and were replaced by a student with the first name Stephan-Lannell, how would this affect the mean?

Need Sorted data from Starting 1-1 "What's in a Name?"

What's the Median for the "Name" Data?

What's the median number of consonants in the first names?

Finding the Median

Organize

- Display the sorted data from Starting 1-1 on the screen:
  STAT (to display the STAT EDIT menu)
  ENTER (to select 1:Edit)
- How many pieces of data are there?
- Is the number of pieces even or odd?
- Use ∨ to scan *x* values (in L_1) to find the median (middle score).

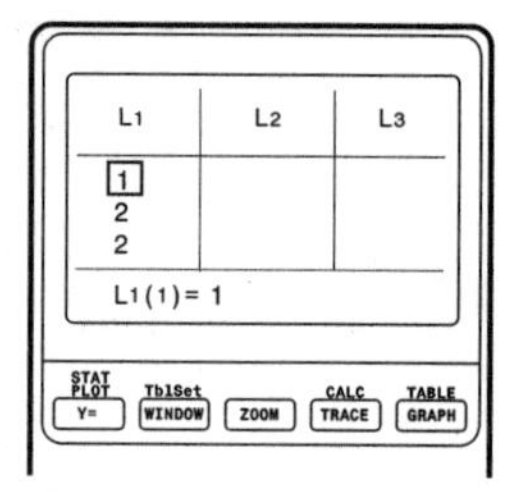

Communicate

- How does the calculator help you find the median?
- What does the median tell you about the data?
- How is finding the median for an "even" data set different from finding it for an "odd" data set?

Reflect

- If the student with the smallest number of consonants left and were replaced by a student with the first name Stephan-Lannell, how would this affect the median?

Need Sorted data from Starting 1-1 "What's in a Name?"

Quartering the "Name" Data

What are the lower and upper quartiles for the "What's in a Name?" data?

Finding the Lower Quartile

Organize

- Display the sorted data from Starting 1-1 on the screen:
  [STAT] (to display the STAT EDIT menu)
  [ENTER] (to select **1:Edit**)
- How many pieces of data are *below* the median (lower half)? Is this number of pieces even or odd?
- Use [v] to scan the *x* values (in L_1) to find the *median of the lower half* of the data.

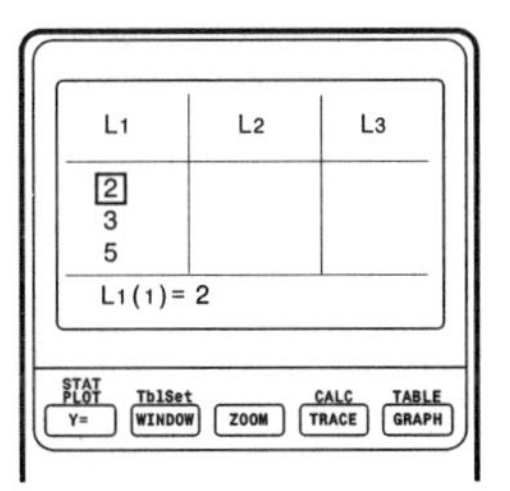

Communicate

- Why is the value you found called the "lower quartile"?
- How would you determine the "upper quartile"? Find it.

Reflect

- If the student with the smallest number of consonants left and were replaced by a student with the first name Stephan-Lannell, how would this affect the lower quartile?

Need Sorted data from Starting 1-1 "What's in a Name?"

What's the Mode for the "Name" Data

What's the mode for the number of consonants in the first names?

Finding the Mode

Organize

- Display the sorted data from Starting 1-1 on the screen:
  [STAT] (to display the STAT EDIT menu)
  [ENTER] (to select **1:Edit**)
- Use [v] to view the data and tally below the number of times each x value (number of consonants) appears.

Number of Consonants in a Name	1	2	3	4	5	6
Tally (Frequency)						

- Find the mode.

Communicate

- What does the mode tell you about the data?
- Will there ever be more than one mode? (Give an example.)

Reflect

- If the student with the smallest number of consonants left and were replaced by a student with the first name Stephan-Lannell, how would this affect the mode?

Need Sorted data from Starting 1-1 "What's in a Name?"

Measuring the Spread of the "Name" Data

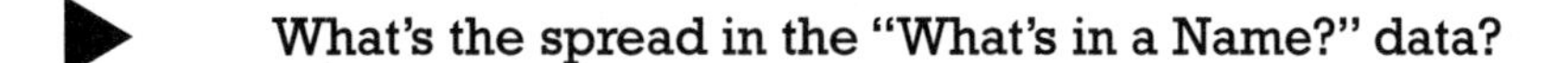

What's the spread in the "What's in a Name?" data?

Finding the Range

Organize

- Display the sorted data from Starting 1-1 on the screen:
  STAT (to display the STAT EDIT menu)
  ENTER (to select **1:Edit**)
- Find the lowest and the highest *x* values, then calculate the difference to find the range.

Communicate

- What does the range tell you about the data?
- How accurate a picture of the spread does it give?

Finding the Interquartile Range (IQR)

Organize

- List the lower and upper quartile values on a piece of paper.
- Calculate the difference to find the interquartile range.

Communicate

- What does the interquartile range tell you about the data?

Reflect

- If the student with the smallest number of consonants left and were replaced by a student with the first name Stephan-Lannell, how would this affect the range? the interquartile range?
- How do the range and interquartile range compare as measures of the spread of the data?

Need Sorted data from Starting 1-1 "What's in a Name?"; values of lower and upper quartiles from Starting 1-5 "Quartering the 'Name' Data"

Who's Far Out?

Are there any "outliers" in the "What's in a Name?" data?

Finding Outliers

Outliers are extreme values. They fall 1.5 * IQR beyond the *upper quartile* or 1.5 * IQR below the *lower quartile.*

Organize

- Calculate the values of the upper and lower fences.
- Examine the data for points outside the two fences:
  STAT (to display the STAT EDIT menu)
  ENTER (to select **1:Edit**)

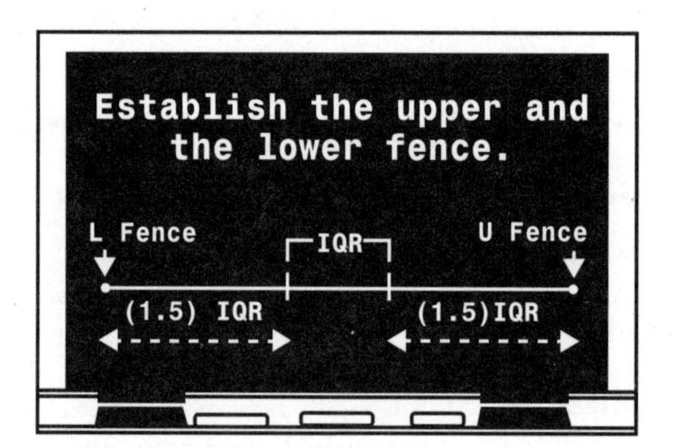

Communicate

- Are there any outliers?
- Describe them.

Reflect

- How could this data set be changed so it would have more than one outlier at the upper end?
- Can this data set have an outlier at the lower end? Why or why not?

Need IQR from Starting 1-7 "Measuring the Spread of the 'Name' Data"; sorted data from Starting 1-1 "What's in a Name?"

Games Won

How could you present the American League data on games won to convey the information effectively?

Making a Stem-and-Leaf Plot

Organize

- Clear the calculator of any previous data (see "Calculator Help 1: Entering Data," page 110).
- Using "Data Set 2: American League/National League Statistics (1993)" (page 131), enter as *x* values in L_1 the number of games won by each team in the American League.
- Sort the data:
  [STAT] (to display STAT EDIT menu)
  2 (to select **2:SortaA**)
  [2nd] [L1]) [ENTER] (to specify L_1 and complete the sort)
- View the sorted data:
  [STAT] [ENTER] (to display STAT EDIT menu and select **1:Edit**)
- Use the sorted data to complete the stem-and-leaf plot below. (Oakland has already been entered.)

GAMES WON IN 1993 BY AMERICAN LEAGUE TEAMS

Stem	Leaf
6	8
7	
8	
9	

Legend: 6|8 = 68

Communicate

- How did the stem-and-leaf plot visually organize the data?
- What does the legend tell you?
- How could you use the stem-and-leaf plot to find the median?

Reflect

- Use "Calculator Program 1: Stem-and-Leaf Plot" (page 125) to generate a stem-and-leaf plot for the data.
- How does the stem-and-leaf plot compare to the one you completed above?

!

Need "Data Set 2: American League/National League Statistics (1993)"; "Calculator Program 1: Stem-and-Leaf Plot"

Runs Scored

How could you present the American League data on runs scored to convey the information effectively?

Making a Stem-and-Leaf Plot

Organize

- Use "Calculator Program 1: Stem-and-Leaf Plot" (page 125).
- Enter "Data Set 2: American League/National League Statistics (1993)" (page 131). (See "Calculator Help 1: Entering Data" page 110.)
- Create a stem-and-leaf plot.
- Examine the plot and determine a legend.

Communicate

- Compare and contrast the stem-and-leaf plot constructed for runs scored to the plot for games won.
- How does the stem-and-leaf plot program handle three-digit data?

Reflect

- Why do you think the program did not use a stem-and-leaf plot such as the one below?

```
6 | 75  84
7 |
8 |
```

Need "Data Set 2: American League/National League Statistics (1993)"; stem-and-leaf plot from Developing 1-1 "Games Won"; "Calculator Program 1: Stem-and-Leaf-Plot"

Truncate or Round?

How could you present the American League data on runs scored in a stem-and-leaf plot by truncating? by rounding?

Organize

Truncating the Data

- Use "Data Set 2: American League/National League Statistics (1993)" (page 131).
- Truncate the data for runs scored. Enter the truncated data in the calculator. (See "Calculator Help 1: Entering Data," page 110.)

RUNS SCORED

	Actual	*Truncated*
Kansas City	675	670
Boston	686	680

- Sort the data if necessary:
  [STAT] (to display STAT EDIT menu)
  2 (to select **2:SortA**)
  [2nd] [L1] [)] [ENTER] (to specify L_1 and complete the sort)
- Use "Calculator Program 1: Stem-and-Leaf Plot" to generate a stem-and-leaf plot for the *truncated* data. Determine a legend.

Communicate

- What does this stem-and-leaf plot tell you?

Rounding the Data

Organize

- Round the data for runs scored. Enter the rounded data in the calculator.

RUNS SCORED

	Actual	*Rounded (nearer 10)*
California	684	680
Detroit	899	900 . . .

- Sort the data if necessary.
- Use "Calculator Program 1: Stem-and-Leaf Plot" to generate a stem-and-leaf plot for the *rounded* data. Determine a legend.

Communicate

- What does this stem-and-leaf plot tell you?

Reflect

- Is the plot for the truncated data less precise than the plot for the rounded data? Explain.
- Compare and contrast these two plots with the one you completed in Developing 1-2.

Need "Data Set 2: American League/ National League Statistics (1993)"; stem-and-leaf plot from Developing 1-2 "Runs Scored"; "Calculator Program 1: Stem-and-Leaf Plot"

Comparing Runs

How could you present the run data for the American League and the run data for the National League to contrast them effectively?

Making a Back-to-Back Stem-and-Leaf Plot

Organize

- In the calculator, enter the runs scored data for both leagues from "Data Set 2: American League/National League Statistics (1993)" (page 131). (See "Calculator Help 1: Entering Data," page 110.)
- Sort the data:
  [STAT] (to display STAT EDIT menu)
  2 (to select **2:SortaA**)
  [2nd] [L1] [)] [ENTER] (to specify L_1 and complete the sort)
- Using the sorted data and on your own paper, make a back-to-back stem-and-leaf plot. Determine the legend.

Communicate

- What does the back-to-back stem-and-leaf plot tell you about the runs scored by the two leagues in 1993?
- Describe the shapes of each distribution.

Reflect

- What are two other sets of data that can be compared in this way?

Need "Data Set 2: American League/National League Statistics (1993)"

Boxing Games Won

What is another way of presenting the American League data on games won?

Making a Box-and-Whisker Plot

Organize

- Using "Data Set 2: American League/National League Statistics (1993)" (page 131), enter as *x* values in L_1 the number of games won by each American League team. (See "Calculator Help 1: Entering Data," page 110.)
- Sort the data:
  [STAT] (to display STAT EDIT menu)
  2 (to select **2:SortaA**)
  [2nd] [L1] [)] [ENTER] (to specify L1 and complete the sort)
- Determine the
  Lower extreme (see Word Bank for help) (LE) ____
  Lower quartile (LQ) ____
  Median ____
  Upper quartile (UQ) ____
  Upper extreme (UE) ____
- Examine the sample box-and-whisker plot and, on your own paper, draw a box-and-whisker plot for the data.

BOX-AND-WHISKER PLOT

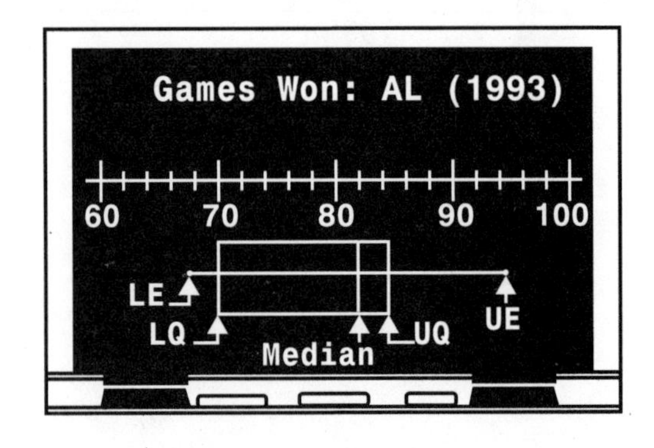

Communicate

- What information is visually presented by the plot?
- Why do you think it is called a "box-and-whisker" plot?
- What percent of the teams won more than 86 games? What percent of the teams won at least 71 games?

Reflect

- Create a box plot for the data using your calculator. Use the following sequence of key presses:
  [2nd] [STAT PLOT] (to display the STAT PLOTS menu)
  1 (to select **1:Plot1 ...**)
  [<] [ENTER] (to turn plot on)
  [V] [>] [>] [ENTER] (to highlight and select the boxplot)
  [V] [ENTER] (to select L_1)
  [V] [ENTER] (to select frequency of 1)
  [GRAPH] (to display box plot)
- How does this box plot compare to the one you drew on paper?

!

Need "Data Set 2: American League/National League Statistics (1993)"

Comparing Games Won

How could you present the games won data for the American League and the National League to contrast them using box plots?

Making Multiple Box Plots

Organize

- Use the calculator to create a box-and-whisker plot for the National League data ("Data Set 2: American League/National League Statistics (1993)," page 131):
  2nd [STAT PLOT] (to display the STAT PLOTS menu)
  V ENTER (to select **1:Plot1 ...**)
  < ENTER (to turn plot on)
  V > > ENTER (to highlight and select the boxplot)
  V > ENTER (to select L2)
  V ENTER (to select frequency of 1)
  GRAPH (to display box plot)

Communicate

- Compare and contrast this box plot with the box plot for the American League data (Developing 1-5 "Boxing Games Won"). Consider medians, quartiles, key statistics, spread, shape, and outliers.

BOX-AND-WHISKER PLOT

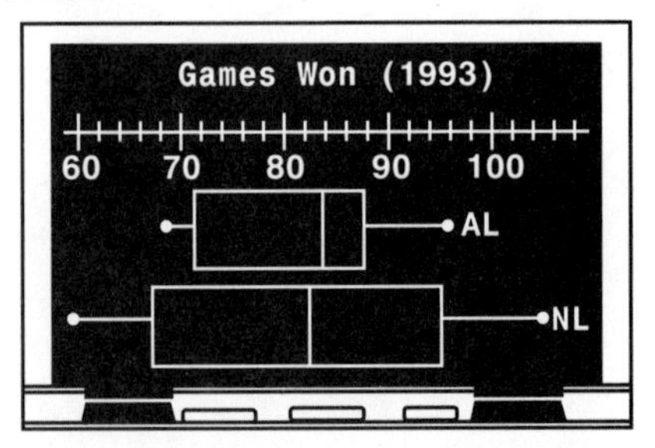

Reflect

- What do the box plots imply about the records of the teams in the two leagues? How might a stem-and-leaf plot help this information?
- If the data from a Japanese League were to be compared with the two United States' leagues, would you use a box plot or a stem-and-leaf plot? Why?

Need Box plot from Developing 1-5 "Boxing Games Won"; "Data Set 2: American League/National League Statistics (1993)"

Runs Scored and Games Won

Is there a relationship between runs scored and games won for the American League data?

Drawing a Scatter Plot

Organize

- Use "Data Set 2: American League/National League Statistics (1993)" (page 131).
- Enter the runs scored for the American League as the *x* variable in (L_1) and games won as the *y* variable in (L_2). (See "Calculator Help 1: Entering Data," page 110.)
- Check the setup:
  [STAT] [>] (to display STAT CALC menu)
  3 (to select **3:SetUp**)
- Draw the scatter plot. (See "Calculator Help 4: Generating a Scatter Plot," page 114.)

Communicate

- Describe the scatter plot.

Reflect

- Which of these best represents the data?

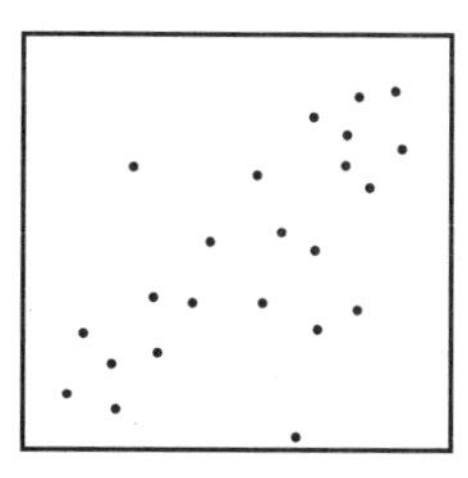

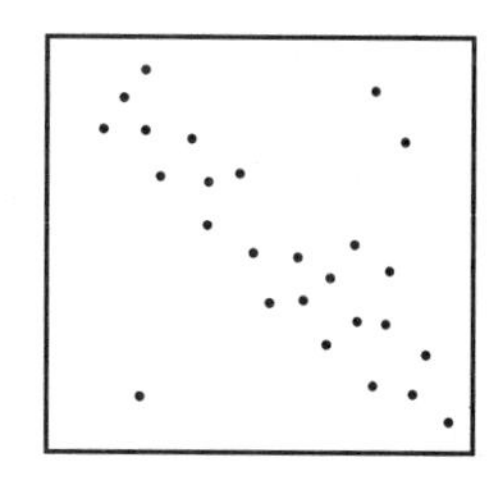

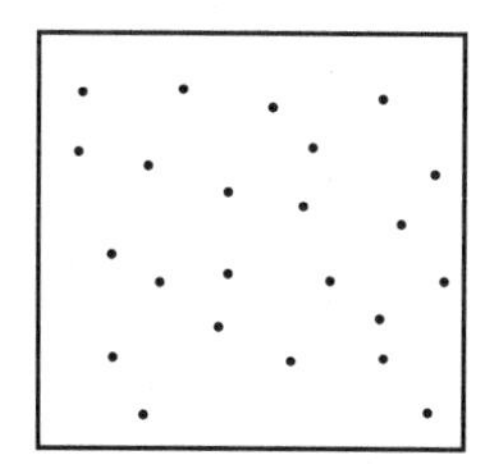

- What does the scatter plot tell you about the data?

Need "Data Set 2: American League/National League Statistics (1993)"

Names and Shoe Lengths

Is there a relationship between the number of consonants in people's first names and their shoe lengths?

Drawing a Scatter Plot

Organize

- Use "Data Set 3: First Names of Students and Their Shoe Lengths" (page 132) or use the first names and shoe lengths (in centimeters) of each of your classmates.
- Enter the number of consonants in the first name as the *x* variable (in L_1) and the shoe length for the same person as the *y* variable (in L_2). (See "Calculator Help 1: Entering Data," page 110.)
- Check the setup:
  STAT > (to display STAT CALC menu)
  3 (to select **3:SetUp**)
- Draw the scatter plot. (See "Calculator Help 4: Generating a Scatter Plot," page 114.)

Communicate

- Describe the scatter plot.
- How does it differ from the scatter plot for the American League data (Developing 1-7)?

Reflect

- Which of these best represents the data?

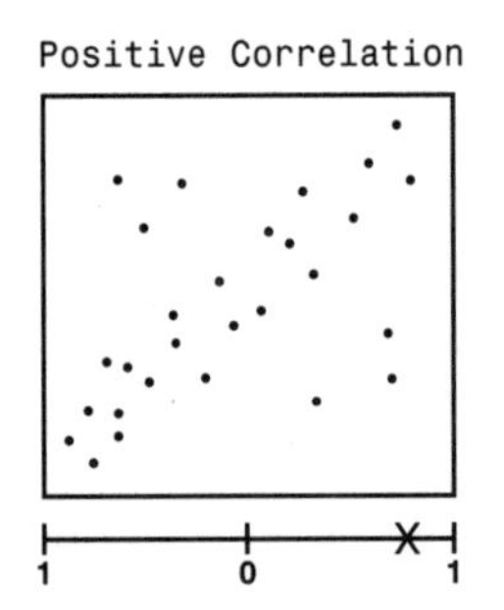

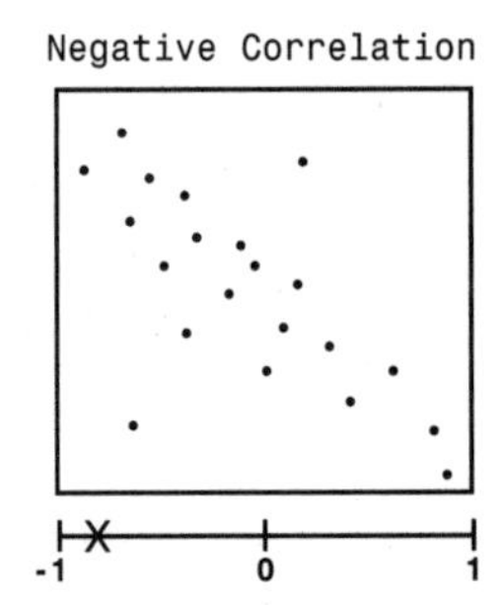

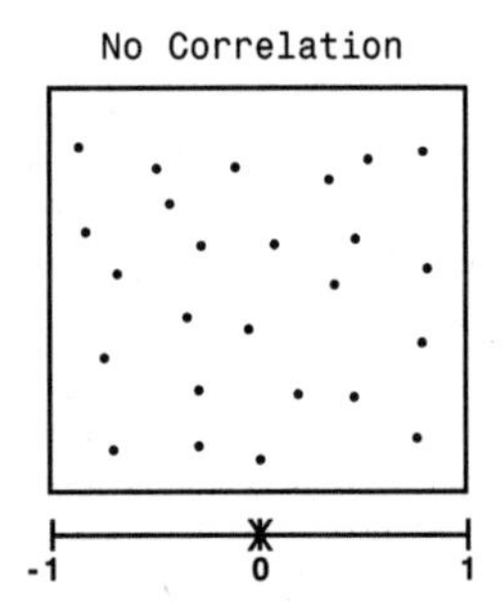

- Where do your data lie on the linear correlation scale?

Need "Data Set 3: First Names of Students and Their Shoe Lengths"; scatter plot from Developing 1-7 "Runs Scored and Games Won"

Roll a Dozen

▶ **What does the sum of 12 rolls of a die tell you?**

How can this information be best displayed?

Displaying and Interpreting the Data

Organize

- Each of you roll the die 12 times and find the sum of the rolls.
- Repeat this experiment ten times (or pool data with classmates.) Enter the ten sums as the *x* variable in L_1 (see "Calculator Help 1: Entering Data," page 110).

Communicate

- Write a brief report with appropriate graphs or plots to describe your data.

Reflect

- What features of the data set led you to display it as you did?
- To what extent does your plot represent all possible sums when a die is rolled 12 times?

!

Need 1 die per pair of students

Who Was the Top Female Tennis Player?

What does the data set on the top women's tennis money winners (1993) tell you?

How can this information be best presented?

Displaying and Interpreting the Data

Organize

- Enter the winnings from "Data Set 4: The Top Women's Tennis Money Winners (1993)" (page 133) as the *x* variable in L_1. (See "Calculator Help 1: Entering Data," page 110.)

Communicate

- Write a brief report with appropriate plots, graphs, or charts that presents and interprets the data.
- What features of the set of data led you to display it as you did?

Reflect

- What if Steffi Graf's earnings were omitted from the data and Jennifer Capriati's earnings of $357,108 were added?
- How would this affect the characteristics of the set of data and the way you would display and interpret it?

Need "Data Set 4: The Top Women's Tennis Money Winners (1993)"

The Top Ten Roadsters

What does the data set on the ten top-selling passenger cars in the United States (1989–1991) tell you?

How can this information be best presented?

Displaying and Interpreting the Data

Organize

- Use "Data Set 5: The Ten Top-Selling Passenger Cars in the United States by Calendar Year (1989–1991)" (page 134).
- Enter the data on the top ten cars. (See "Calculator Help 1: Entering Data," page 110.)

Communicate

- Write a brief report with appropriate plots, graphs, or charts that compares and interprets the data.

Reflect

- What if only the top six were used?
- How would this affect the characteristics of the data set and the way you would display and interpret it?

Need "Data Set 5: The Ten Top-Selling Passenger Cars in the United States by Calendar Year (1989–1991)"

The Speeds of Animals

What does the data set on the maximum speeds attained by animals tell you?

Displaying and Interpreting the Data

Organize

- Enter the speeds (mph) from "Data Set 6: The Speeds of Animals" (page 135) as the *x* variable in L_1. (See "Calculator Help 1: Entering Data," page 110).

Communicate

- Write a brief report that presents and interprets the data. Use appropriate charts, graphs, or plots.

Reflect

- What features of the set of data led you to display it as you did?

Need "Data Set 6: The Speeds of Animals"

Endangered Animals

What does the data set on United States' and foreign endangered animals tell you?

How can this information be best contrasted?

Displaying and Interpreting the Data

Organize

- Use "Data Set 7: United States List of Endangered Species" (page 136). Enter the data on the endangered animals. (See "Calculator Help 1: Entering Data," page 110.)

Communicate

- Write a brief report that compares and interprets the data. Use appropriate graphs, plots, or charts.

Reflect

- What would happen if only the mammals, birds, reptiles, and amphibians were used?
- How would this affect the way you displayed and interpreted the data?

!

Need "Data Set 7: United States List of Endangered Species"

Looking Back

The key concepts in this module are listed below. Using appropriate descriptions, illustrations, or examples, summarize two or more of these in such a way that you could explain them to others in your class.

Measures of Central Tendency
Measures of Dispersion

Organization and Presentation of Data
Histogram
Stem-and-Leaf Plot

Box-and-Whiskers Plot
Scatter Plot

Module 2

Probability, Simulations, and Random Numbers

Looking Ahead

Probabilities, such as a 50 percent chance of getting a "head" when tossing a coin, are based on the symmetry, geometry, or other physical characteristics of probability devices. Such probabilities are called *theoretical probabilities.*

While theoretical probabilities are important in mathematics, experimental probabilities are often used to make predictions in our daily life—especially in areas such as surveys and weather predictions. In order to determine an experimental probability, you need to carry out a number of trials of a random experiment, record the data, then analyze and interpret the data. Once experimental probabilities have been determined, you can solve many complex problems.

Flip the Coin

If a coin is flipped, how could it land?

On the scale below mark your probability prediction for a head with an *X*.

0	$\frac{1}{2}$	1
Impossible		Certain

Displaying the Experimental Probability

Organize

- Working with a partner, flip the coin 20 times and record the number of heads.
- Repeat this experiment nine more times.
- Enter the data (see "Calculator Help 1: Entering Data," page 110). Put x values into L_1 and y values into L_2. The x values are 0, 1, 2, . . ., 20—that is, the number of times heads occurs. The y values are the corresponding frequencies. For example: $x_7 = 6$, $y_7 = 2$ means six heads occurred twice.
- Use the [WINDOW] key to check the range and then make the histogram (see "Calculator Help 2: Range; Histogram," page 111).
- Calculate the mean. (See "Calculator Help 3: Mean," page 113):
  [STAT] [>] (to display the STAT CALC menu)
  [ENTER] [ENTER] (to display **1-Var Stats**)
- Read, and *record* $\bar{x}$ for later use.

Communicate

- From the histogram display, which number(s) of heads occurred most frequently?
- Complete and discuss the information in the table.

	heads	*tails*
$\bar{x}$		
$\frac{\bar{x}}{20}$		

Reflect

- If the experimental probability for heads is $\frac{\text{\# of heads}}{\text{\# of flips}}$, why does $\frac{\bar{x}}{20}$ represent the experimental probability using the class data?
- Why does the data set from ten experiments provide a better probability estimate than data from one experiment?

Need 1 coin for each pair of students

Model the Coin Flip

For a fair coin what would you assume are the probabilities for a head? a tail? Why are they equally likely?

Using Probabilities

Organize

- Use the random-number generator on your TI-82. (See "Calculator Help 5: Using the Random-Number Generator," page 115.)

Communicate

- If the numbers 0, 1, 2, 3, and 4 each represent a head, and the numbers 5, 6, 7, 8, and 9 each represent a tail, how do the ten numbers produced by the random-number generator model ten tosses of a coin?
- List the string of numbers as a sequence of head-tail outcomes.

Reflect

- We assumed each outcome was equally likely. Contrast this with the experimental probabilities in Starting 2-1.
- Why do you think these assumed probabilities are called "theoretical" probabilities?

Need Experimental probabilities from Starting 2-1 "Flip the Coin"

Modeling Two Flips

When a fair coin is flipped what do you assume is the probability of a head?

What do you predict is the probability of getting two heads on two consecutive flips? Mark it on the scale with an *X*.

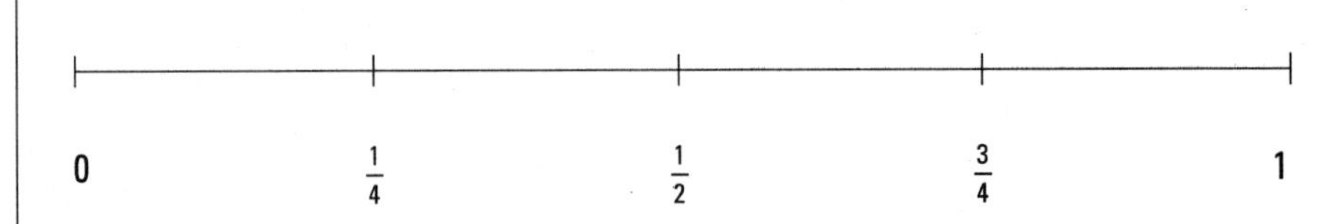

Modeling the Coin Flips

Organize

- Use the random-number generator (see "Calculator Help 5: Using the Random-Number Generator," page 115).
- If the numbers 0, 1, 2, 3, and 4 each represent a head, and the numbers 5, 6, 7, 8, and 9 each represent a tail, use randomly-generated pairs to model two consecutive flips.
- Continue until you have 30 pairs on your list.

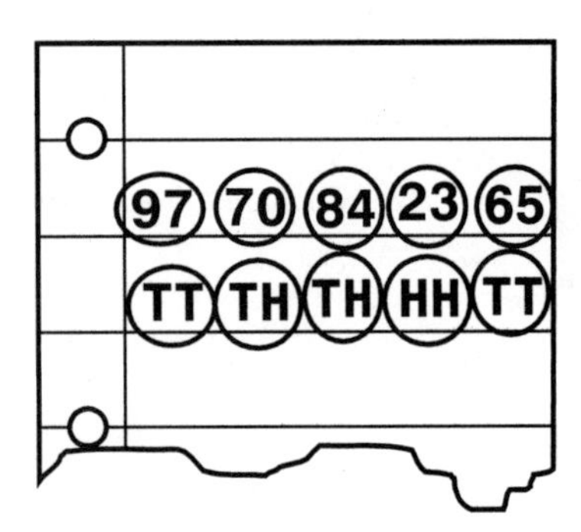

Communicate

- How many pairs have two heads?
- What is the experimental probability of two heads?

  Experimental Probability = $\frac{\text{number of successes}}{\text{total possibilities (30)}}$

- Write about it. How did you use the random-number generator to help you find the probability of getting two heads?

Reflect

- If you did another 30 trials do you think the probability would be the same?
- How else could heads and tails be assigned to the digits?

Steffi's Serve

Steffi is ready to serve. Given her serve statistics, what do you estimate are her chances of a double fault? Mark it on the scale with an *X*.

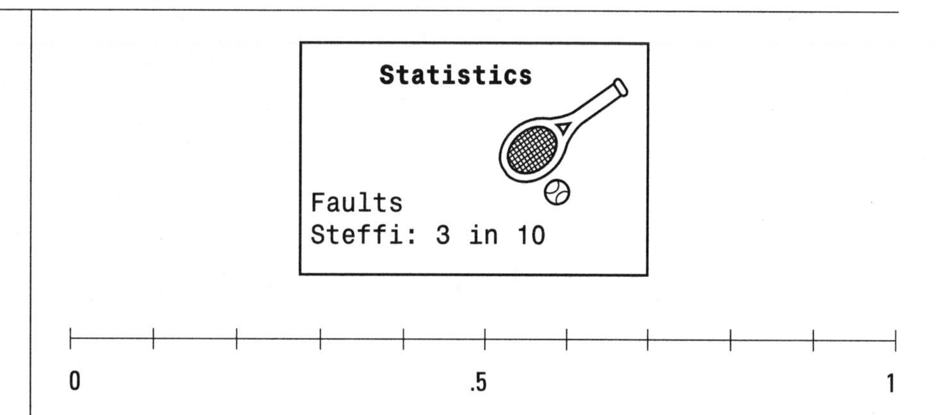

Modeling the Serves

Organize

- Use the random-number generator (see "Calculator Help 5: Using the Random-Number Generator," page 115) to call up multiple sets of ten random numbers.
- Using the relevant probabilities, decide which single-digit numbers represent "fault" and "good serve."
- Model two serves by grouping consecutive pairs of digits.
- Did digits that represent a fault come up twice in a pair (double fault) or not? Tally.
  "Double Fault" ____ "Good Serve" ____
- Repeat until you have made 50 tallies.

Communicate

- How many double faults were there?
- What was the total number of possibilities?
- What is the experimental probability of a double fault?

  $$\text{Experimental Probability} = \frac{\text{number of successes}}{\text{total possibilities}}$$

Reflect

- About how often does Steffi serve a double fault?
- Was your prediction about right?
- What is the chance of a double fault on a "bad" day when Steffi serves six faults out of ten?

Flip the Cup

If a cup is flipped, how could it land?

Mark with an *X* and label on the scale your probabilities for each outcome.

0	$\frac{1}{2}$	1
Impossible		Certain

Determining Experimental Probability

Organize

- Working with a partner, flip the cup 20 times and record the result of each outcome.
- Repeat this experiment 11 times.
- Enter the data (see "Calculator Help 1: Entering Data," page 110). The *x* values are 0, 1, 2, . . ., 20—that is, the number of times "cup up" occurs—and *y* values are the corresponding frequencies. For example, $x_5 = 6$, $y_5 = 4$ means six "cup up" results occurred in four experiments.
- Calculate the mean. (See "Calculator Help 3: Mean," page 113):
  [STAT] [>] (to display STAT CALC menu)
  [ENTER] [ENTER] (to display **1-Var**)
- Read, and *record* $\bar{x}$ for later use.
- CLEAR AND REPEAT FOR THE OTHER 2 OUTCOMES.

Communicate

- Complete and discuss the information in the table.

	Up	*Down*	*Sideways*
$\bar{x}$			
$\frac{\bar{x}}{20}$			

Reflect

- Why does $\frac{\bar{x}}{20}$ represent the experimental probability?
- How does the experimental probability for the cup differ from those for the coin?

Need 1 paper cup for each pair of students; probabilities from Starting 2-1 "Flip the Coin"

Chances for the Name

If a first name is randomly selected for your class, what is the chance that it has four consonants?

Determining Experimental Probability

Organize

- Use "Data Set 1: First Names of Students" (page 130) or the first names of students in your class. Enter the data (see "Calculator Help 1: Entering Data," page 110).

  The *x* values (L_1) are 1, 2, 3, . . .—that is, the number of consonants in the first names. The *y* values (L_2) are the corresponding frequencies. For example, $x_4 = 4$, $y_4 = 8$ means the number of students with four consonants in their first name is eight.

- Use the WINDOW key to check the range and then make the histogram (see "Calculator Help 2: Range; Histogram," page 111).
- Determine the number of students with four consonants in their names.

Communicate

- What is the theoretical probability that a name randomly drawn will have four consonants? Explain your thinking.
- How could you use the histogram to find the theoretical probability of drawing a five-consonant name?

Reflect

- What is the sum of the probabilities of all possible outcomes for this experiment?
- If Stephan-Lannell joined the group, how would this affect the sum of the probabilities?

Need "Data Set 1: First Names of Students"; consonant data from Starting 1-1 "What's in a Name?"

Modeling the Chances for the Name

▶ **Why does this spinner model the experiment for flipping a fair coin?**

How could you construct a spinner to model the experiment for the number of consonants in first names?

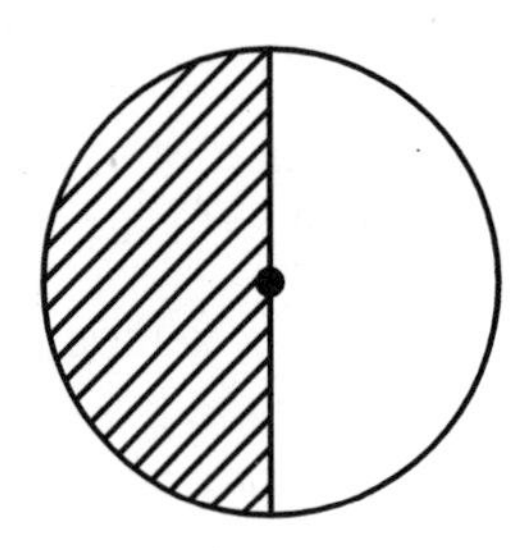

Modeling Names

Organize

- Using the theoretical probabilities on consonants in classmates' first names (Starting 2-6), construct a circle graph.

Communicate

- How does your circle graph represent the probabilities of each of the outcomes in the Starting 2-6 experiment?
- How do you know when all outcomes have been accounted for in your circle spinner?

Reflect

- How would your spinner change if Stephan-Lannell joined the class?
- Would a spinner for a different class be identical to the one for this class? Explain.

Need Experimental probabilities from Starting 2-6 "Chances for the Name"

Two Broadway Favorites

Examine "Data Set 8: Broadway Favorites" (page 137). If a name were drawn at random what do you predict is the probability that the person has seen both Broadway shows?

Modeling Shows Seen

Organize

- Using "Data Set 8: Broadway Favorites," tally to complete a matrix such as this one:

	Seen *Les Miserables*	Not Seen *Les Miserables*
Seen *Phantom of the Opera*		
Not Seen *Phantom of the Opera*		

Communicate

- Using the matrix, what percentage of students have seen both shows? neither show? *Phantom*? *Les Miserables*?
- What are the probabilities that a student drawn at random has seen:
  Phantom ____ *Both Shows* ____
  Les Miserables ____ *Neither Show* ____

Reflect

- How many probabilities could you have determined from the matrix? Describe each using labels from the matrix.

!

Need "Data Set 8: Broadway Favorites"

Three in a Row

Using the information from Starting 2-8:

- **What is the probability that a student drawn at random has seen both Broadway shows?**
- **What do you predict is the probability that three consecutive draws (with replacement) will result in students who have seen both shows?**

Modeling Three Draws

Organize

- Use the random-number generator to call up multiple sets of ten random numbers (see "Calculator Help 5: Using the Random-Number Generator," page 115).
- Using the relevant probabilities, decide which single-digit numbers represent "seen both" and "not seen both."
- Model three draws by grouping consecutive sets of three digits.

  Did digits that represent "seen both" come up three times in a group or not? Tally.
  "Three Times" ____ "Not Three Times" ____
- Repeat until you have made 50 tallies.

Communicate

- How many times did "seen both" come up three times?
- What were the total number of possibilities?
- What is the experimental probability of getting three "seen both" results in three consecutive draws?

Reflect

- Suppose the experimental probability of drawing a student who had seen both shows was 0.685. How would this affect the assignment of random numbers?

Need Experimental probabilities from Starting 2-8 "Two Broadway Favorites"

Three Hits

Al Bazuka has a batting average of .325. Assuming his probability of a hit remains constant, what is the probability of three consecutive hits in a game?

Modeling Three Bats

Organize

- Use the random-number generator to call up multiple sets of ten random numbers (see "Calculator Help 5: Using the Random-Number Generator," page 115).
- Using the relevant probabilities, decide which three-digit numbers represent "a hit" and "not a hit."
- Model three times at bat by grouping consecutive sets of nine digits. Did digits that represent "a hit" come up three times in a group or not? Tally.
  "Three Hits" ____ "Not Three Hits" ____
- Repeat until you have made 50 tallies.

Communicate

- How many times did a hit come up three times in a row?
- What is the experimental probability of Al making three consecutive hits in a row?

Reflect

- If the question had been to determine the probability that Al made four consecutive hits in a game, how would this have affected the assignment of random numbers and the number of strings needed?

The Top Ten

Use "Data Set 9: Pop Albums" (page 138) or collect your own top-ten data.

What album has been on top most often during the first ten weeks of the year?

How many times in the ten weeks has it been on top?

Based on its record what is the probability that it will be on top in the eleventh week?

What do you estimate is the probability that it will be on top at least two of the next three weeks?

Simulating Three Weeks

Organize

- Use the random-number generator to call up multiple sets of ten random numbers (see "Calculator Help 5: Using the Random-Number Generator," page 115).
- Using the relevant probabilities, decide which single-digit numbers represent "on top" and "not on top."
- Model three weeks by grouping consecutive sets of three digits. Did digits that represent "on top" come up at least two times in a group or not? Tally.
  "At Least Two on Top" ____ "Less Than Two on Top" ____
- Repeat until you have made 50 tallies.

Communicate

- How many times did the album come up at least two times in three weeks?
- What is the experimental probability of the album coming up at least two times in the next three weeks?

Reflect

- If the question had been to find the probability that the album comes up at *most* two times in the next three weeks, how would this have affected your simulation?

Need "Data Set 9: Pop Albums"

On Top for Four Weeks

Use the information from "Data Set 9: Pop Albums" (page 138). What do you estimate is the probability that *Music Box* will be on top for the next four weeks?

Simulating Four Weeks

Organize

- Use the random-number generator to call up multiple sets of ten random numbers (see "Calculator Help 5: Using the Random-Number Generator," page 115).
- Using the relevant probabilities, decide which single-digit numbers represent "on top" and "not on top."
- Model four weeks by grouping consecutive sets of four digits.
- Did digits that represent "on top" come up four times in a group or not? Tally.
  "Four Times on Top" ____ "Not Four Times on Top" ____
- Repeat until you have made 50 tallies.

Communicate

- How many times did the album come up four times in four weeks?
- What is the experimental probability of the album coming up four times in the next four weeks?

Reflect

- If the question had been to find the probability that the album comes up at *least* three times in the next four weeks, how would this have affected your simulation?

Need "Data Set 9: Pop Albums"

Basketball Playoffs

In basketball playoffs the two finalists play a series until one team wins four games. What is the maximum number of games that could be played?

What do you estimate is the probability that a series between two *evenly matched* teams will go seven games?

Simulating the Series

Organize

- Use the random-number generator to call up multiple sets of ten random numbers (see "Calculator Help 5: Using the Random-Number Generator," page 115).
- Given that the teams are evenly matched, decide which single-digit numbers represent a "win" or a "loss" for one team.
- Model a seven-game series. Did it go seven games or not? Tally.
  "Seven Games" ____ "Not Seven Games" ____
- Repeat until you have made 50 tallies.

Communicate

- How many times did it take seven games to complete the series?
- What is the experimental probability that the series will go to seven games?

Reflect

- If the question had been to determine the probability that the series would last at most seven games, how would this have affected your simulation?
- What is the probability that the series would have lasted at most seven games? How can you determine this without carrying out a simulation?

Grand-Slam Match

In a grand-slam tennis final the two finalists play a series of sets until one player wins three sets. What is the maximum number of sets that could be played?

If the odds in favor of Pete beating Jim in any set are 3:2, estimate the probability that the match will not be completed in three straight sets.

Modeling a Grand-Slam Match

Organize

- Use the random-number generator to call up multiple sets of ten random numbers (see "Calculator Help 5: Using the Random-Number Generator," page 115).
- Model a grand-slam match. Did it go three sets or not? Tally.
  "Three Sets" ____ "Not Three Sets" ____
- Repeat until you have made 50 tallies.

Communicate

- How many times did the match last for more than three sets?
- What is the experimental probability that the match is not completed in three sets?

Reflect

- If the question had been to determine the probability that the match was not completed in three sets *and* was won by Pete, how would this have affected your simulation?

Chasing the Circle

If you randomly prick a circle mat 20 times, how often do you predict that it will land in or on a circle?

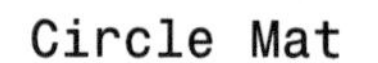

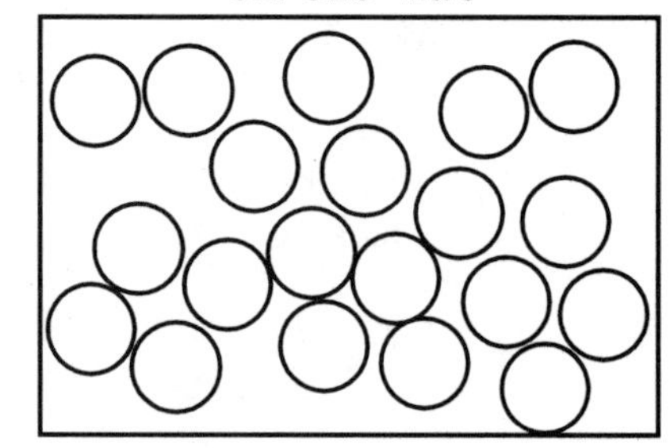

Finding the Probability of Hitting a Circle

Organize

- Use "Blackline Master 1: Circle Mats" (page 104) and a pin. With eyes closed, "prick" the top Circle Mat 20 times with the pin.
- Count how many times you hit a circle.
- Repeat with your other mat.
- Pool your data with nine other students and complete a table such as the following:

Number of Circles Pricked	*Frequency (Number of Students with This Many)*
0	
1	
2	
...	
19	
20	

- Enter the "Number of Circles Pricked" as the x value in L_1 and the "Frequency" as the y value in L_2 (see "Calculator Help 1: Entering Data," page 110).
- Find the mean (see "Calculator Help 3: Mean," page 113).

Communicate

- What does the mean of your group represent?
- Use the group mean to find the experimental probability that the pin lands in or on a circle.

Reflect

- How did your experimental probability compare with that from the pooled data?
- Which is the better estimate? Why?

!

Need a straight pin and "Blackline Master 1: Circle Mats" for each pair of students

Estimating π

How could you use the data from Developing 2-6 to find the area of each of the circles on a Circle Mat?

Finding the Areas

Organize

- Measure the diameter of a circle in millimeters and calculate the total area of the 20 circles as ____ π sq. mm.
- Measure the dimensions of the card in millimeters and calculate its area in square millimeters.

Communicate

- How is the ratio of the $\frac{\text{(total area of circles)}}{\text{(total area of the card)}}$ related to the class experimental probability found in Developing 2-6?

Reflect

- How could you use the ratio of the areas above and the experimental probability to find an estimate of π?
- Find the estimate for π using the pooled data.
- How close is your estimate?

Need Circle Mats and group data from Developing 2-6 "Chasing the Circle"; millimeter ruler for each pair of students

Bazuka Bats

Al Bazuka has a batting average of .325. In his next game he will bat four times. What do you estimate is the probability that he will make zero hits? at least two hits? four hits?

Simulating Bazuka at Bat

Organize

- Use the random-number generator (see "Calculator Help 5: Using the Random-Number Generator," page 115) and an appropriate grouping to simulate Al's four times at bat.

Communicate

- What is the experimental probability that Al had:
  zero hits?
  at least two hits?
  four hits?
- Contrast these experimental probabilities with the theoretical probabilities.

Reflect

- If you had to redetermine Al's batting average after each time at bat, how would this affect your simulation process?

Rock-Star Cards

Each box of Korny Krisps has one rock-star card inside. If there are five different rock-star cards evenly distributed among the boxes, predict how many boxes of cereal you would expect to buy to get all five cards.

Simulating the Rock-Star Cards

Organize

- Use the random-number generator (see "Calculator Help 5: Using the Random-Number Generator," page 115) and designate each two digits as a different rock-star card.
- Model opening a sequence of boxes until you get all five cards and record the results.
- Repeat an appropriate number of times.
- Use "Calculator Help 1: Entering Data" (page 110) and "Calculator Help 3: Mean" (page 113) to find the average number of boxes needed to get all five cards.

Communicate

- Describe your process and explain why it works.

Reflect

- Besides using random numbers, how else could you have simulated this problem? Explain.
- If there were seven different rock-star cards, how would your simulation process change?

Raining in Surfers' Paradise

In Surfers' Paradise the probability that a rainy day is followed by a rainy day is 0.6; the probability that a sunny day is followed by a rainy day is 0.8.

If Sunday is sunny, predict the experimental probability that there will be exactly three rainy days this week.

Simulating the Weather for a Week

Organize

- Use the probabilities above to find the probability that a rainy day is followed by a sunny day and the probability that a sunny day is followed by a sunny day.
- Use the random-number generator (see "Calculator Help 5: Using the Random-Number Generator," page 115) to model the weather for a week, starting with a sunny day on Sunday. Did it have exactly three rainy days?
- Repeat an appropriate number of times to determine the experimental probability of having exactly three rainy days in a week.

Communicate

- Describe and explain the process you used to determine the experimental probability of having exactly three rainy days in a week.

Reflect

- How would your simulation have changed if Sunday had been rainy?
- Besides random numbers how else could you have simulated this problem? Explain.

Looking Back

The key concepts in this module are listed below. Using appropriate descriptions, illustrations, or examples, summarize two or more of these in such a way that you could explain them to others in your class.

Experimental Probability
Theoretical Probability

Geometrical Probability

Using Random Numbers to Model and Simulate
Random Numbers

Module 3

Counting

Looking Ahead

Many problems in our daily lives involve complex counting or determinations of probabilities that are based on complex counting. A good example of the latter is determining the chances of a person winning a state lottery by picking the correct six numbers.

To determine the probability of winning a lottery, you need to find the number of ways of selecting 6 numbers from 54—a problem that is extremely difficult, especially if you try to list the individual sets of 6 numbers. In this module, you will explore various ways of counting to solve real-life problems such as this.

Naming the Dog

The Evert family is getting a new dog. Each of the three children thinks up six different names for the dog, and the parents each suggest four names. If no name appears on more than one list, how many names were suggested for the dog?

Solving the Problem

Organize

- Model the problem.
- Make a list of names that might have been suggested by:

  —the children

  —the parents
- There should be no duplications.
- Determine how many names were suggested.
- If you were to do it again would it be the same number?

Communicate

- Describe how you solved the problem.
- Would there be any way to solve the problem more easily? Explain.

Reflect

- A pet shop donated a dog to the Hanover School ball team for their mascot. Students voted to give the dog a name. There were three classes of 27 students and four classes of 23 students. Each student suggested four names. If all the names were different, how many names were suggested?

How Many Faces?

A red die and a green die are rolled, and their sum is found.

How many outcomes have sums that are greater than seven or less than seven?

How many outcomes have even sums or sums less than six?

Solving the Problem

Organize

- Work with another student.
- Complete the table to list all possible sums.

Green \ Red	1	2	3	4	5	6
1	2	3				
2	3					
3						
4						
5						
6						

- Find the solutions to the questions above.

Communicate

- Compare your results with those of other students.
- Describe how the strategies for solving the two problems were different. Why were they different?

Reflect

- There is a contest at the annual carnival that involves spinning two wheels with the numbers 1 through 88 on each.

  1. How many different outcomes are possible?
  2. How many outcomes have odd sums or even sums?
  3. You win a prize if you spin a one on the first wheel or a one on the second wheel. How many different ways can you win?

Birth Month in a Box

The names and birth months of all the students in a class are put in a box. How many names must be drawn to be certain of having two people with the same birthday month?

Solving the Problem

Organize

- Cut apart the cards on "Blackline Master 2: Names and Birth Months" on page 105. Place them in a box.
- Draw names at random out of the box. Count the number of draws until two people with the same birth month have been drawn.
- Repeat two more times.

Communicate

- Did the number of draws vary? Why?
- For the worst possible case how many draws would be needed to be certain of getting two students with the same birthday month?
- Suppose it was known that no one in the class had a birthday in October or November. How many draws would be needed to get two alike in this case?

Reflect

- As part of a leap-year celebration each student in a school wrote down his or her name and birthday on a piece of paper and turned it in. The principal noted that every day in the year was represented. If the names are put in a box, how many must be drawn to be *certain* of having two people with the same birthday?

Need "Blackline Master 2: Names and Birth Months"

Three the Same

The names and birth months of 52 students are put in a box. How many names must be drawn to be certain of having 3 students with the same birthday month? 4 students? 5 students?

Solving the Problem

Organize

- Cut apart the cards on "Blackline Master 3: More Names and Birth Months" (page 106) and combine them with the cards from "Blackline Master 2: Names and Birth Months" (page 105). Put all the names in a box.
- Draw names at random out of the box and count the number of draws until three people with the same birth month have been drawn.
- Repeat two more times.

Communicate

- Compare your results with those of other students. Did results vary? Why?
- For the *worst* possible case how many draws would be needed to be certain of getting three students with the same birthday month?
- Deduce the result for four students having the same birthday month; then do the same for five students.

Reflect

- Generalize your results for the case of n students having the same birthday month.

Need "Blackline Master 2: Names and Birth Months"; "Blackline Master 3: More Names and Birth Months"

First Names in a Box

In a class of 20 students there are 3 Juans, 2 Zacs, 2 Kims, and 4 Eds. If all the first names in this class are put in a box, how many names must be drawn to be certain of getting two students with the same first name?

Solving the Problem

Organize

- Work with another student. Model the problem.
- Repeat several times. For the worst possible case how many names must be drawn to be certain of getting two people with the same first name?

Communicate

- Compare your results with those of other students.
- Describe why your solution works for the worst possible case.

Reflect

- What if 3 Mikes joined the class of 20 above? Would this change the solution? Explain.

How Many Sandwiches?

If a school lunch menu offered sandwiches with three choices of meat and two choices of bread, how many different kinds of sandwiches could you buy?

Solving the Problem

Organize
- Model the problem. Make an organized list of the choices.

Communicate
- Compare your results and strategies with those of other students.

Reflect
- How could the result be obtained in a simpler way?
- To help celebrate the holidays the school cooks provided a special menu. There were five choices for the main dish, six choices for vegetables, three choices for salad dressings, four choices for fruit, six choices for desserts, and three choices for drinks. How many different meals could be chosen?

How Many Kinds of Pizza?

If only one topping is allowed for the special price on Sam's Pizza Menu, how many different kinds of pizza can be selected?

Sam's Pizza Menu		
Toppings		
Sausage		Pepperoni
Canadian Bacon		Ham
Crusts		
Thick		Thin
Sizes		
Small	Medium	Large

Solving the Problem

Organize

- Model the problem. If you selected sausage as the topping, how many kinds of pizza could you get? List them.
- How many different kinds of pizza would there be for each of the following:
  sausage? ____ ham? ____
  pepperoni? ____ Canadian bacon? ____

Communicate

- How many different kinds of pizza in all? Explain your reasoning.

Reflect

- How could you have worked out the number of different kinds of pizza without listing them?
- Suggest a general rule and apply it to the following problem. At Super Sam's Pizza there are:
  14 meat toppings
  12 vegetable/fruit toppings
  3 kinds of cheese
  4 sizes
  2 crusts
- How many kinds of pizzas are at Super Sam's when you choose one item from each category?

Batting Order

Amy, Bev, and Carol are the first three batters on the softball team, but they often bat in a different order. How many ways of ordering them in the first three places are there?

Solving the Problem

Organize

Solve the problem in the two following ways:

- List all the different orders.
- Consider, in order, how many ways each batting position can be filled.

First Position	*Second Position*	*Third Position*

Communicate

- If $n!$ (called n factorial) $= n * (n-1) * (n-2) \ldots 2 * 1$, describe your solution to the batting order problem using factorial notation.

Reflect

- If the entire softball team has nine players and any batting order is allowed, how many different batting orders are possible?

Pitcher Bats Last

On a baseball team with nine players, the manager decides that any batting order is allowed as long as the pitcher bats last. How many different batting orders are possible?

Solving the Problem

Organize

- Solve the problem and describe your answer using factorial notation: ________________!
- Evaluate ____! using the calculator. For example, to find 6!, follow this sequence of key presses:
  6 (enter number)
  [MATH] [<] (to display the MATH PRB menu)
  4 (to select **4:!**)
  [ENTER] (to get 720 as the answer)

Communicate

- Did you expect there to be so many different batting orders?
- What would you predict for the number of possible batting orders for an eleven-person cricket team with no restrictions? Check it with your calculator.

Reflect

- On a baseball team with nine players, the catcher and the pitcher can only bat in one of the last two positions. How many different batting orders are possible? Write your answer in factorial notation, then evaluate it using the calculator.

Raffle Tickets

At a small ice cream social, raffle tickets are assigned a three-digit number in the following way:
Each digit is a 0, 1, 2, or 3.
No two digits are alike.

How many different raffle tickets are there?

Solving the Problem

Organize

- Model the problem. If the first digit is 0, list each of the possible raffle tickets.

	Number of Second Digits	*Third Digits*
0		

Communicate

- How many different raffle tickets would there be where the first digit is 0? 1? 2? 3?

Reflect

- How many different raffle tickets are there? Explain your reasoning.
- How could you have worked out the number of different raffle tickets without listing them?
- Suggest a general rule and apply it to the problem where all ten digits can be used to form the three-digit numbers (with no digit repeated).

Calculating the Number of Raffle Tickets

At another ice cream social, raffle tickets are assigned a four-digit number in the following way:

Each digit is 0, 1, 2, 3, 4, or 5.

No two digits are alike.

How many different raffle tickets are there?

Solving the Problem

Organize

- Solve the problem. Use the diagram to help you. Express the answer as a product.

Number of First Digits	*Number of Second Digits*	*Number of Third Digits*	*Number of Fourth Digits*

- Use your calculator to determine the number of arrangements of six digits taken four at a time:
  6 (enter number)
  [MATH] [<] (to display MATH PRB menu)
  2 (to select **2:nPr**)
  4 [ENTER] (to complete the computation)

Communicate

- Compare your solution to that found with the calculator. Why do you think they are the same?
- 6 $_nP_r$ 4 is usually written $_6P_4$. Describe what you think $_6P_4$ means and relate this to the raffle problem.

Reflect

- Construct a problem for which the answer is $_{10}P_4$. Explain why your problem works and find the value of $_{10}P_4$.

"Special" Raffle Tickets

Raffle tickets at a fair are assigned a five-digit number in the following way:

Each digit is a 0, 1, 2, 3, 4, 5, 6, 7, 8, or 9.
No two digits are alike.
The first digit cannot be a 0.

How many different raffle tickets are there?

Solving the Problem

Number of First Digits	*Number of Second Digits*	*Number of Third Digits*	*Number of Fourth Digits*	*Number of Fifth Digits*

Organize

- Solve the problem using the diagram to help. Express your solution using ${}_nP_r$ notation.
- Use your calculator to evaluate it.

Communicate

- Describe how your completed diagram and your ${}_nP_r$ solution are related.

Reflect

- How many *more* raffle tickets would there be if zero were allowed in the first position?

"Extra-Special" Raffle Tickets

Raffle tickets at a fair are assigned a five-digit number in the following way:

Each digit is a 0, 1, 2, 3, 4, 5, 6, 7, 8, or 9.
No two digits are alike.
Neither the first nor the second digit can be a 0.

How many different raffle tickets are there?

Solving the Problem

Number of First Digits	*Number of Second Digits*	*Number of Third Digits*	*Number of Fourth Digits*	*Number of Fifth Digits*

Organize

- Solve the problem using the diagram to help. Express your solution using ${}_nP_r$ notation.
- Use your calculator to evaluate it.

Communicate

- Describe how your completed diagram and your ${}_nP_r$ solution are related.

Reflect

- Construct a problem for which the answer is $3!({}_9P_3)({}_7P_2)$. (Hint: Try modifying the above problem.)
- Explain why your problem works and find the value of $3!({}_9P_3)({}_7P_2)$.

Mini-Lotto

In Mini-Lotto you pick two different digits from the set 0, 1, 2, and 3. How many different pairs could be selected?

Solving the Problem

Organize

- Solve a simpler problem: How many different two-digit numbers can be formed from 0, 1, 2, and 3 where no digit is repeated? List them.
- Express your answer in $_nP_r$ notation and evaluate it.
- Now solve the Mini-Lotto problem. (Hint: If the digits 1 and 2 were drawn, tickets "1-2" and "2-1" would both be winners since order does not matter.)

Communicate

- Describe how you would modify your list for the simpler problem to find the listing of pairs for Mini-Lotto.

Reflect

- Compare your Mini-Lotto solution to the solution for the simpler problem. Explain how they are related.

Calculating the Mini-Lotto Pairs

In another Mini-Lotto you pick two different digits from the set 1, 2, 3, and 4. How many different pairs could be selected?

Solving the Problem

Organize

- Create and solve the simpler problem, as in Developing 3-7.
- Express your answer in ${}_nP_r$ notation and evaluate it.
- Now solve the Mini-Lotto problem. (Remember, order is not important.)
- Use your calculator to address this problem:
  4 (enter number)
  MATH < (to display MATH PRB menu)
  3 (to select **3:nCr**)
  2 ENTER (to complete the calculation)

Communicate

- Compare your solution to the Mini-Lotto problem to that found by the calculator. Why do you think they are the same?
- 4 ${}_nC_r$ 2 is usually written ${}_4C_2$. Describe what you think ${}_4C_2$ means and relate this to the Mini-Lotto problem itself.

Reflect

- Compare ${}_4C_2$ and ${}_4P_2$. How are they related? Interpret this in terms of the simpler problem and the Mini-Lotto problem itself.

Pick Three

In Pick Three mini-lotto you pick three different digits from the set 1, 2, 3, and 4. How many different sets of three could be selected?

Solving the Problem

Organize

- Solve the simpler problem. How many different three-digit numbers can be formed from 1, 2, 3, and 4 where no digit is repeated? List them.

- Now use your list to solve the Pick Three problem. (Remember, order does not matter.)

Communicate

- If the three winning numbers are 1, 2, and 3, how many different ways could they have been drawn? Explain.
- How does this relate to 3!?

Reflect

- Compare your Pick Three solution to the solution for the simpler problem. Explain how they are related.

Pick Three Again

In Pick Three mini-lotto you pick three different digits from the set 1, 2, 3, and 4. How many different sets of three could be selected?

Solving the Problem

Organize

- Create and solve the simpler problem, as in Developing 3-9.

- Express your answer in ${}_nP_r$ notation and evaluate it.
- Now solve the Pick Three Again problem using the calculator:
  4 (enter number)
  MATH < (to display MATH PRB menu)
  3 (to select **3:nCr**)
  3 ENTER (to complete computation)
- Express your answer in ${}_nC_r$ notation.

Communicate

- Describe how your solution in ${}_nC_r$ notation relates to the Pick Three problem.
- If the three winning numbers are 2, 3, and 4, how many different ways could they have been drawn? Explain how this relates to 3!

Reflect

- How are ${}_4C_3$ and ${}_4P_3$ related?
- Can you generalize a relationship for ${}_nC_r$ and ${}_nP_r$?

Many the Same

All the students in a school wrote their names and birthdays on paper and turned them in. The principal noted that every day was represented except the twenty-ninth of February. If the names are put in a box how many names must be drawn to be certain of having 2, 3, 4, . . . *n* students with the same birthday?

Solving the Birthday Problem

Organize

- Model the problem. Make an organized list of your results for each of the cases.

Communicate

- Describe the general result for *n* students having the same birthday and explain how you found it.

Reflect

- What if one day, for example February 2, occurred only once as a birthday? How would this affect your results?
- How many students would there need to be in the school to guarantee that five students have their birthday on the same date?
- Are there any other limitations that could affect the general solution? Explain.

Two in the Same Month

Suppose five students are selected at random from a large school enrollment. What is the probability that at least two of them have the same birthday month (not necessarily the same day or year)?

Solving the Birth Month Problem

Organize

- How many ways with no restrictions are there of selecting birthday months for the five students? Try completing the table:

First Student	*Second Student*	*Third Student*	*Fourth Student*	*Fifth Student*
12	12			

- Use your calculator to determine the result.
- How many ways of selecting the five birthday months are there if no two students can have the same birthday month? Express your answer in appropriate notation (${}_nP_r$ or ${}_nC_r$) and determine the value.

Communicate

- Use your work above. Describe how to find the number of different ways of selecting the five birthday months if at least two students must have the *same* birthday month.

Reflect

- What is the probability that at least two of the students have the same birthday month? Explain.

The State Lottery

In one state lottery, players choose 6 numbers from 54. If only one group of 6 numbers wins the prize, what is the probability that you will win if you purchase only one set of 6 numbers?

Solving the State Lottery Problem

Organize

- How many ways are there of selecting 6 different numbers from 54? Express your answer in appropriate notation(${}_nP_r$ or ${}_nC_r$) and determine the number of ways.
- How many ways are there of selecting the winning set of 6?

Communicate

- Describe how you could use your work above to find the probability of winning the state lottery if you purchase only one set of 6 numbers.

Reflect

- Would the result be changed if you purchased two sets of 6 numbers?
- In the Australian Gold Lotto, players select 6 numbers out of 45. How does the strategy you used for the State Lottery problem enable you to determine the probability of winning the Australian Gold Lotto?

Winning Hands

In a game of poker a player is dealt 5 cards at random from a pack of 52. If there are no second draws what is the probability of getting a full house (three of a kind and two of a kind)?

Solving the Problem

Organize

- How many ways are there of selecting five different cards from a regular deck? Express your answer in appropriate notation and determine the number of ways.
- How many ways are there of drawing a full house?

Communicate

- Describe how you could use your work above to find the probability of getting a full house.

Reflect

- What are the chances of getting four of a kind?
- What are the chances of getting a flush (five cards in the same suit)?

Looking Back

The key concepts in this module are listed below. Using appropriate descriptions, illustrations, or examples, summarize two or more of these in such a way that you could explain them to others in your class.

Addition Rule
Multiplication Rule

Pigeonhole Principle

Permutations and Factorial Notation
Combinations

Module 4

Modeling and Predicting

Looking Ahead

Many phenomena in our world generate important data and require predictions to be made based on these data. For example, environmental scientists are concerned about rising ocean water levels during the last 100 years and the effect this increase might have on our coastlines and urban areas.

To address this problem they have collected data on water levels over an extended period and have found that a linear or straight-line relationship exists between year and water level. Using this straight line they can make predictions about water levels into the twenty-first century. During this unit you will investigate both linear and nonlinear data arising from various real-world settings in order to make useful predictions from data.

Shoe and Arm Lengths

If you know someone's shoe length is 25 cm, what would you predict for their arm length in centimeters?

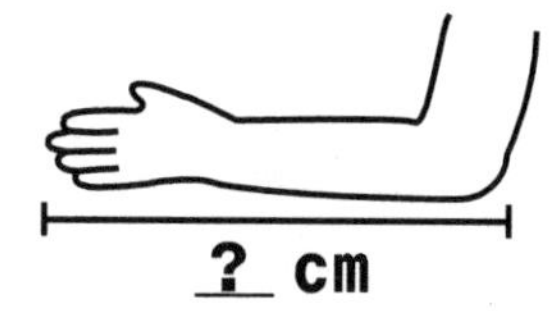

Gathering and Displaying the Data

Organize

- Use "Data Set 3: First Names of Students and Their Shoe Lengths" (page 132) or collect your own data on shoe and arm length to create a table such as the following:

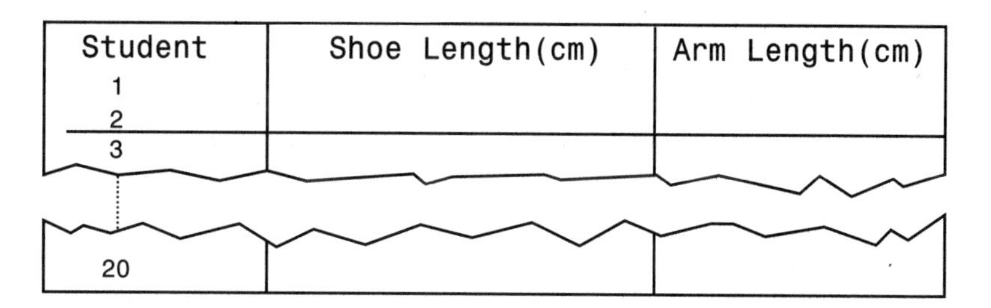

Student	Shoe Length(cm)	Arm Length(cm)
1		
2		
3		
20		

- Enter the data using shoe lengths as the x values in L_1 and arm lengths as the y values in L_2 (see "Calculator Help 1: Entering Data," page 110).
- Check the setup:
  STAT > (to display STAT CALC menu)
  3 (to select **3:SetUp**)
- Set the range (see "Calculator Help 2: Range," page 111).
- Draw the scatter plot (see "Calculator Help 4: Generating a Scatter Plot," page 114).

Communicate

- What does the scatter plot suggest about the points?
- Use the line function (see "Calculator Help 6: Using the Line Function," page 116) to draw in the line you think best describes the data. Explain your strategy.
- Are there any points that do not fit well? Comment.

Reflect

- If a new student entered your class with a shoe length of 35 cm, what does your line predict for the student's arm length? (See "Calculator Help 6: Cursor Movement in the Line Function," page 116.)
- Would your prediction be the same as other students in your class? Why?

Need "Data Set 3: First Names of Students and Their Shoe Lengths"

Shoe and Name Lengths

What is the relationship between a student's shoe length and the number of consonants in his or her first name?

Gathering and Displaying the Data

Organize

- Use "Data Set 3: First Names of Students and Their Shoe Lengths" (page 132) or collect your own data on shoe and name length.
- Enter the shoe lengths as the *x* variable in L_1 and the number of consonants in the first name as the *y* variable in L_2 (see "Calculator Help 1: Entering Data," page 110).
- Check the setup:
  [STAT] [>] (to display the STAT CALC menu)
  3 (to select **3:SetUp**)
- Set the range (see "Calculator Help 2: Range," page 111).
- Draw the scatter plot (see "Calculator Help 4: Generating a Scatter Plot," page 114).

Communicate

- How is the scatter plot for this data set different from that for Starting 4-1 "Shoe and Arm Length"?
- Use the line function (see "Calculator Help 6: Using the Line Function," page 116) to draw in the line you think best describes the shoe length-name length data.
- Are there many points that do not fit well? Explain.

Reflect

- Why was it more difficult to fit a line to the shoe length-name length data than to the data in Starting 4-1?

Need One calculator in each group showing the scatter plot from Starting 4-1 "Shoe and Arm Length"

The Missing Movie

What is the relationship between money made by a movie and its rank order?

Displaying the Movie Data

Organize

- Use "Data Set 10: Top Money-Making Movies of All Time," page 139. Enter a movie's rank order as the *x* variable in L_1 and the money it made as the *y* variable in L_2 (see "Calculator Help 1: Entering Data," page 110). Note that you cannot enter data for *The Empire Strikes Back.*
- Check the setup:
  [STAT] [>] (to display STAT CALC menu)
  3 (to select **3:SetUp**)
- Set the range (see "Calculator Help 2: Range," page 111).
- Draw the scatter plot (see "Calculator Help 4: Generating a Scatter Plot," page 114). Instead of [ZOOM] **9** push [GRAPH].

Communicate

- Describe the scatter plot.
- Draw a line that best describes the movie data on rank order and money made (see "Calculator Help 6: Using the Line Function," page 116).
- Are there many points that do not fit well? Explain.

Reflect

- Knowing that *The Empire Strikes Back* ranks fifth on the list, how much money do you predict it made? (See "Calculator Help 6: Using the Line Function," page 116).

Need "Data Set 10: Top Money-Making Movies of All Time"

Shoe Lengths and Consonants

How well does a straight line describe the relationship between students' shoe lengths and the number of consonants in their first names?

Gathering and Displaying the Data

Organize

- Use "Data Set 3: First Names of Students and Their Shoe Lengths" (page 132) and only information on the first five students to complete the following table:

Student	*x = Shoe Length*	*y = Number of Consonants*
1		
2		
3		
4		
5		

Communicate

- Find the mean ($\bar{x}$) of the shoe lengths and the mean ($\bar{y}$) of the number of consonants. (See "Calculator Help 3: Mean," page 113.)
- On graph paper construct a scatter plot of the data. Plot and label the point ($\bar{x}$, $\bar{y}$). (You may use "Blackline Master 4: Quarter-Inch Graph Paper," page 107.)
- Keeping the piece of spaghetti on ($\bar{x}$, $\bar{y}$), rotate it until you get the line you think best describes your data. Draw this line on your scatter plot.

Reflect

- Describe how you fitted the line to your data.
- What is the greatest vertical distance of any point from the line of best fit ($|y - y'|$)?
- Interpret ($|y - y'|$) for each point of your scatter plot.

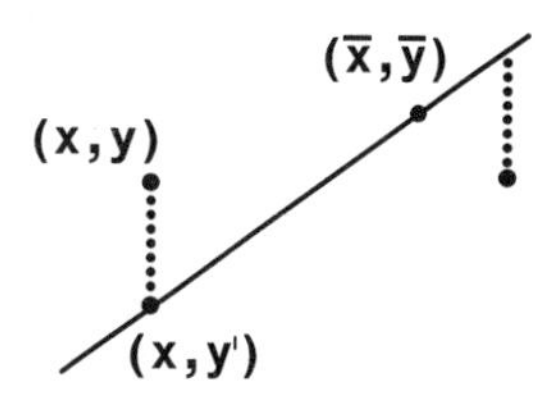

!

Need "Blackline Master 4: Quarter-Inch Graph Paper"; dry spaghetti or lengths of thread

The Line of Best Fit

How can we draw the line that best describes the relationship between students' shoe lengths and the number of consonants in their first names?

Fitting a Line to Data

Organize

- Use the scatter plot and the line you generated in Developing 4-1 "Shoe Lengths and Consonants":
- Determine the slope and the y-intercept of your line and find its equation in the form $y' = mx + b$.

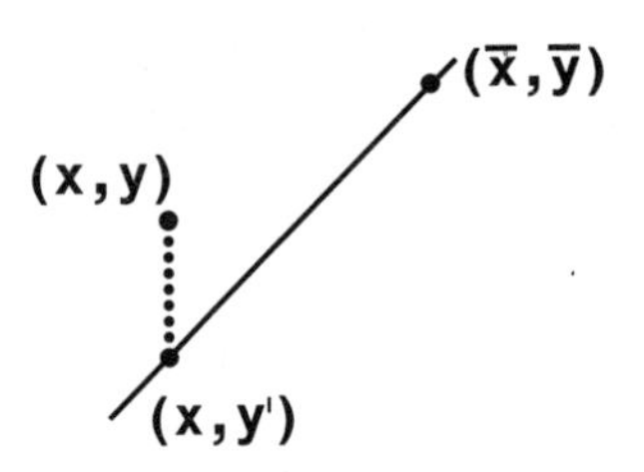

- Find $(|y - y'|)$ for each point on the scatter plot, and then find the SUM of all $(y - y')^2$.
- Note: y' represents the y coordinates of points on your line of best fit. (See diagram.)
- Finally, find the average $\frac{\Sigma(y - y')^2}{n}$, where n is the number of points on the scatter plot. This average is called the *mean square error.*

Communicate

- Describe what you did in finding the mean square error.
- Interpret what it means geometrically.
- Compare your mean square error with those of other students.

Reflect

- Which students have the least mean square error?
- Is this the line of "best fit"? Why?

!

Need Scatter plot and line generated in Developing 4-1 "Shoe Lengths and Consonants"

Equation for the Line of Best Fit

For the shoe length and number of consonants data, how can the calculator produce the line with the least mean square error?

Finding the Equation

Organize

- Use "Data Set 3: First Names of Students and Their Shoe Lengths" (page 132). Enter shoe length as the x variable in L_1 and the number of consonants in a student's first name as the y variable in L_2 (see "Calculator Help 1: Entering Data," page 110).
- Check the setup:
  STAT > (to display the STAT CALC menu)
  3 (to select **3:SetUp**)
- Check the range (see "Calculator Help 2: Range," page 111).
- Draw the scatter plot (see "Calculator Help 4: Generating a Scatter Plot," page 114).
- Find the y-intercept b and the slope a of the line of best fit (see "Calculator Help 7: Finding the y-Intercept, the Slope of the Line of Best Fit, and the Correlation Coefficient (r)" page 117).
- Complete the equation:
  $y = \_\_ x + \_\_$
  $(y = ax + b)$

Communicate

- How do the slope and intercept of this line compare with the line having the least mean square error? (See the *Reflect* section of Developing 4-2.)
- Why would you expect the slopes and intercepts of the two lines to be similar?

Reflect

- Which of the two lines above would have the least square error? Why?

Need "Data Set 3: First Names of Students and Their Shoe Lengths"; information from Developing 4-2 "The Line of Best Fit"

Graphing on the Scatter Plot

For the shoe length and number of consonants data, how can you use the equation of the line of best fit to plot the line on the scatter plot?

Using the Equation

Organize

- Draw the scatter plot (see "Calculator Help 4: Generating a Scatter Plot," page 114).
- Draw the line of best fit (see "Calculator Help 7: Finding the *y*-Intercept, the Slope of the Line of Best Fit, . . . Using the Equation to Draw the Line of Best Fit," page 117).

Communicate

- Describe how close the line "fits" the points.
- Why can we not do any better for this data set?

Reflect

- Put an *X* on the scale to indicate the correlation between shoe length and number of consonants in the first name. Explain your response.

Strong Negative Correlation	No Correlation	Strong Positive Correlation
–1	0	1

Need In the calculator: data from shoe length and number of consonants in the first name; intercept b and slope a

Finding the Correlation

What is your prediction for the correlation between shoe length and number of consonants in the first name?

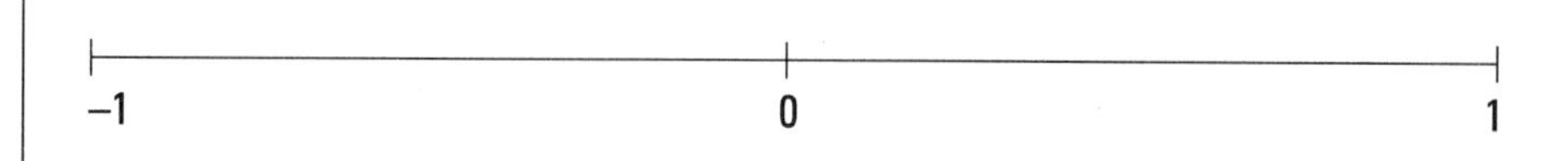

Using the Calculator

Organize

- Find the correlation coefficient (r) (see "Calculator Help 7: Finding the y-Intercept, the Slope of the Line of Best Fit, and the Correlation Coefficient (r)," page 117).

Communicate

- Describe the strength and direction of the correlation.
- Contrast this with how close the line of best fit was to the points.

Reflect

- What is the relationship between correlation and "closeness" of the line of best fit?
- How would you describe this data set in terms of both the correlation and the "closeness" of the line of best fit?

Need "Data Set 3: First Names of Students and Their Shoe Lengths"

Money Made and Movie Ranking

Based on the movie data on rankings (x variable) and money made (y variable), what is your prediction for the correlation?

Strong Negative	No	Strong Positive
−1	0	1

Finding the Correlation

Organize

- Use "Data Set 10: Top Money-Making Movies of All Time," page 139.
- Enter the data (see "Calculator Help 1: Entering Data," page 110).
- Find the correlation coefficient (r) (see "Calculator Help 7: Finding the y-Intercept, the Slope of the Line of Best Fit, and the Correlation Coefficient (r)," page 117).

Communicate

- Describe how strong the correlation is between rankings (x) and money made (y).
- As the ranking numbers get larger, what happens to money made on a movie?
- How is this reflected in the correlation value?

Reflect

- How close would the line of best fit be to the data points? Why do you think so?
- Check your thinking:
- Draw the scatter plot (see "Calculator Help 4: Generating a Scatter Plot," page 114).
- Draw the line of best fit (see "Calculator Help 7: Finding the y-Intercept, the Slope of the Line of Best Fit, . . . Using the Equation to Draw the Line of Best Fit," page 117).

!

Need "Data Set 10: Top Money-Making Movies of All Time"

Money Made and Year Released

▶ **Based on the movie data on year released (x variable) and money made (y variable), what is your prediction for the correlation?**

Strong Negative	No	Strong Positive
–1	0	1

Finding the Correlation

Organize

- Enter the data (see "Calculator Help 1: Entering Data," page 110).
- Find the correlation coefficient (r) (see "Calculator Help 7: Finding the y-Intercept, the Slope of the Line of Best Fit, and the Correlation Coefficient (r)," page 117).

Communicate

- Describe how strong the correlation is between year released (x) and money made (y).
- As the years increase, what happens to the money made?
- How is this reflected in the correlation value?

Reflect

- Compare the two relationships:
  movie ranking—money made
  year released—money made
- How were the correlation coefficients
  alike?
  different?

Need "Data Set 10: Top Money-Making Movies of All Time"

Forecasting the Missing Movie

For the movie data on year released (x variable) and money made (y variable), what is your prediction for the money made on *The Empire Strikes Back* (1980)?

Finding the Standing

Organize

- Find the y-intercept b and slope a for the line of best fit (see "Calculator Help 7: Finding the y-Intercept, the Slope of the Line of Best Fit, and the Correlation Coefficient (r)," page 117).
- Complete the equation:
  $y =$ __ $x +$ __
  ($y = ax + b$)
- Use the equation to forecast the money made (y) when the year released (x) = 1980 (see "Calculator Help 8: Using the Best-Fit Equation to Predict a y Value for a Given x," page 118).

Communicate

- Interpret the y value you obtained. How would you describe it to a movie fan?

Reflect

- Based on the line of best fit, how much money would you expect this year's top movie to make?
- What factors might affect the accuracy of this prediction?

Need In the calculator: "Data Set 10: Top Money-Making Movies of All Time"

Shoe Lengths and Arm Lengths

What would you predict for a person's arm length, if their shoe length is 28 cm?

Gathering and Displaying Data

Organize

- Use "Data Set 3: First Names of Students and Their Shoe Lengths" (page 132). Enter shoe length as the *x* variable and arm length as the *y* variable (see "Calculator Help 1: Entering Data," page 110).
- Draw a scatter plot to display the relationship between shoe length and arm length (see "Calculator Help 4: Generating a Scatter Plot," page 114).
- Use your calculator to find the line of best fit for the data (see "Calculator Help 7: Finding the *y*-Intercept, the Slope of the Line of Best Fit, . . . Using the Equation to Draw the Line of Best Fit," page 117).
- Find the arm length corresponding to a shoe length of 28 cm (see "Calculator Help 8: Using the Best-Fit Equation to Predict a *y* Value for a Given *x*," page 118).

Communicate

- Describe what you did to determine the arm length for a 28-cm shoe length.
- How good is your prediction? Justify.

Reflect

- Would your line of best fit be just as valid for determining arm length for a student whose shoe length is 10 cm? Explain your thinking.

Need "Data Set 3: First Names of Students and Their Shoe Lengths"

Testing the Radius and Area Relationship

Would you predict a linear relationship between the radius of a circle and its area?

Testing Linearity

Organize

- Determine the radius and area of each circle on "Blackline Master 5: Circles," page 108.
- Enter the data using radius as the x variable and area as the y variable (see "Calculator Help 1: Entering Data," page 110).
- Draw the scatter plot (see "Calculator Help 4: Generating a Scatter Plot," page 114).
- Use your calculator to find any other measure that will help you test for linearity.

Communicate

- What did your scatter plot and other measure(s) suggest about the linearity of the data? Explain your reasoning.

Reflect

- Why should you have expected data involving the radius and area of a circle to be nonlinear?

Need "Blackline Master 5: Circles"

Transforming the Radius and Area Data

▶ **How would you describe the shape of the scatter plot of the data in Extending 4-2?**

How could you transform the data set so it could be fitted by a straight line?

Testing Linearity for the Transformed Data

Organize

- Define a new variable, y_1, where $y_1 = \sqrt{y}$

- Enter the data for x and y_1 (see "Calculator Help 1: Entering Data," page 110).
- Draw the scatter plot for the (x,y_1) data (see "Calculator Help 4: Generating a Scatter Plot," page 114).
- Use your calculator to find any other measure that will help you test for linearity.

Communicate

- What did your new scatter plot and other measure(s) suggest about the linearity of the (x,y_1) data? Explain your reasoning and compare the linearity measures of the (x,y_1) data and the (x,y) data.

Reflect

- Is this consistent with the mathematics in the think bubble above? Explain.

Need Scatter plot from Extending 4-2 "Testing the Radius and Area Relationship"

Fitting Radius to Area

How could you use the (x,y_1) data to predict the area when the radius is 1.5?

Finding the Line of Best Fit

Organize

- Use the calculator to find the line of best fit for the (x,y_1) data (see "Calculator Help 7: Finding the *y*-Intercept, the Slope of the Line of Best Fit, . . . Using the Equation to Draw the Line of Best Fit," page 117).
- Write the equation in the form: $y_1 = ax + b$
- Use this equation and the transformation $y_1 = \sqrt{y}$ to find an equation (†) relating radius (x) to area (y).

Communicate

- Use the $y_1 - x$ equation $y_1 = ax + b$ to find y_1 when $x = 1.5$. Then find y when $x = 1.5$.
- Use $y - x$ equation (†) to find y when $x = 1.5$.

Reflect

- Do the two processes above lead to the same value for area when the radius is 1.5? Why or why not?

Need In the calculator: (x,y_1) data from Extending 4-3 "Transforming the Radius and Area Data"

Testing the Radius and Volume Relationship

Would you predict a linear relationship between the radius of a sphere and its volume?

Finding a "Best-Fit" Relationship

Organize

- Determine the radius (x) and volume (y) of at least five spheres.
- Use an appropriate transformation to predict the volume of a sphere when the radius is 1.5.

Communicate

- Describe the process you used to predict the volume of a sphere with radius 1.5 from your transformed data.
- Compare the process with that used to predict the area of a circle from its radius.

Reflect

- Compare your predicted value of the volume of a sphere of radius 1.5 with the exact value determined from the standard formula ($v = (\frac{4}{3})\ \pi r^3$). Why would you have expected this result?

Crude-Oil Production

Would you predict a linear relationship between year and number of barrels of crude oil produced worldwide?

Testing Linearity

Organize

- Use "Data Set 11: Annual World Crude-Oil Production (1980–1992)," page 140. Enter the data using year as the *x* variable and number of barrels of crude oil produced as the *y* variable (see "Calculator Help 1: Entering Data," page 110).
- Draw the scatter plot (see "Calculator Help 4: Generating a Scatter Plot," page 114).
- Use your calculator to find any other measure that will help you test for linearity.

Communicate

- What did your scatter plot and other measure(s) suggest about the linearity of the data? Explain your reasoning.

Reflect

- Could the data be fitted by a nonlinear function?

Need "Data Set 11: Annual World Crude-Oil Production (1980–1992)"

Transforming the Crude-Oil Production Data

How would you describe the shape of the scatter plot for the data in Extending 4-6?

How could you transform the data set so it could be better fitted by a straight line?

Testing Linearity for the Transformed Data

Organize

- Use "Data Set 11: Annual World Crude-Oil Production (1880–1992)," page 140.
- Define a new variable x_1 where $x_1 = \ln x$.

- Enter the x_1 data in L_3 (see "Calculator Help 9: Entering Transformed Data," page 119).
- Draw the scatter plot for the (x_1,y) data (see "Calculator Help 4: Generating a Scatter Plot," page 114).
- Use your calculator to find any other measure that will help you test for linearity.

Communicate

- What did your new scatter plot and other measure(s) suggest about the linearity of the (x_1,y) data? Explain your reasoning and compare the linearity measures for the (x_1,y) data and the (x,y) data.

Reflect

- Would the transformed data (x_1,y) be a better predictor of crude-oil production in the year 2000 than the original data? Explain.

Need In the calculator: "Data Set 11: Annual World Crude-Oil Production (1880–1992)"

Fitting Year and Crude-Oil Production

How could you use the (x_1,y) data to predict the number of barrels of crude oil produced in the year 2000?

Finding the Best-Fit Equation

Organize

- Use the calculator to find the slope and y-intercept for the (x_1,y) data (see "Calculator Help 7: Finding the y-Intercept, the Slope of the Line of Best Fit, and the Correlation Coefficient (r)," page 117).
- Write the best-fit linear equation in the form: $y = ax_1 + b$.
- Use this equation and the transformation $x_1 = \ln x$ to show that the best-fit equation relating year (x) to number of barrels produced (y) is $y = a(\ln x) + b$, where a and b are the numbers found in $y = ax_1 + b$.
- Use the equation relating y to x_1 to predict y when $x = 2000$. (What is x_1 when $x = 2000$?)

Communicate

- Use the equation relating y to x to predict y when $x = 2000$.

Reflect

- Do the two processes above lead to the same value for "number of barrels produced" in 2000? Justify your answer.

Need In the calculator: the (x,y_1) data from Extending 4-7 "Transforming the Crude-Oil Production Data"

An Alternative Transformation

Could the scatter plot for the Extending 4-6 data be more exponential than logarithmic?

How could you transform the data to fit an exponential relation?

Testing Linearity for the Exponential Transformation

Organize

- Use the think bubble to define a new variable $y_1 = \ln y$

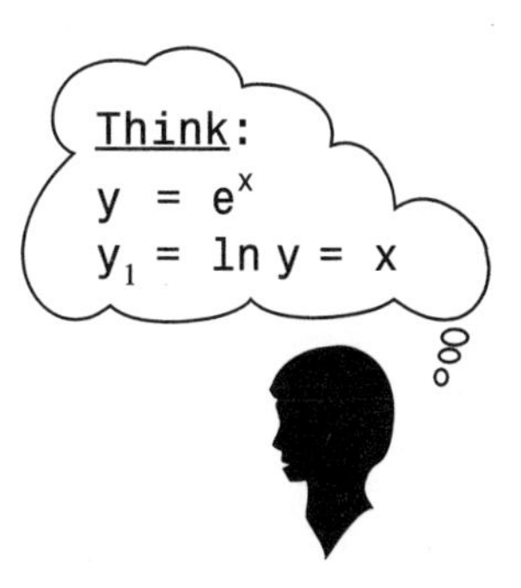

- Enter the data for y_1 in L_4 (see "Calculator Help 9: Entering Transformed Data," page 119).
- Draw the scatter plot and check for linearity (see "Calculator Help 4: Generating a Scatter Plot," page 114).
- Find the best fit linear equation in the form: $y_1 = ax + b$.
- Use the transformation $y_1 = \ln y$ to find the best-fit equation relating y to x.

Communicate

- Describe the process you followed and use the best-fit equation relating y to x to predict the number of barrels in the year 2000.

Reflect

- Compare this transformation with that used in Extending 4-8. How did you determine which transformation produced the "best fit"? Explain.

Need In the calculator: "Data Set 11: Annual World Crude-Oil Production (1880–1992)"; information from Extending 4-8 "Fitting Year and Crude-Oil Production"

Another Transformation

Could the scatter plot for the Extending 4-6 data be better described by a power function rather than an exponential or logarithmic function?

How could you transform the data to fit a power function?

Testing Linearity for the Power Transformation

Organize

- Use the think bubble to define new variables y_1 and x_1.

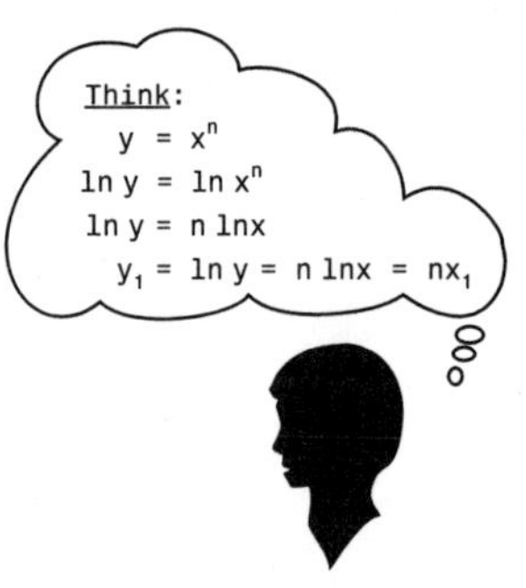

- Enter the (x_1,y_1) data (see "Calculator Help 1: Entering Data," page 110).
- Draw the scatter plot (see "Calculator Help 4: Generating a Scatter Plot," page 114), test for linearity, and find the best-fit equation in the form $y_1 = ax_1 + b$ (see "Calculator Help 7: Finding the y-Intercept, the Slope of the Line of Best Fit, and the Correlation Coefficient (r)," page 117).
- Then find the best-fit equation relating y to x.

Communicate

- Describe the process you followed and use the best-fit equation relating y to x to predict the number of barrels in the year 2000.

Reflect

- Compare this transformation with those used in Extending 4-8 and Extending 4-9. How would you determine which of the three transformations produced the "best fit"? Explain.
- Is the fit provided by the best of these transformations better than the linear fit using the original data? Explain.

Need In the calculator: "Data Set 11: Annual World Crude-Oil Production (1880–1992)"; information from Extending 4-8 "Fitting Year and Crude-Oil Production"; information from Extending 4-9 "An Alternative Transformation"

Non-English-Speaking Americans

What is the relationship between the number of non–English-speaking Americans in 1980 and the number in 1990?

Finding a Relationship

Organize

- Use "Data Set 12: Number of Non–English-Speaking Americans," page 141. Enter 1980 data as the *x* variable and 1990 data as the *y* variable (see "Calculator Help 1: Entering Data," page 110).
- Test various transformations and decide which best represent or model the data.
- Determine one or more "best-fit" equations.

Communicate

- Describe the process you used to model the data. Can more than one model be justified? Explain.

Reflect

- Use your model(s) to predict the Spanish-speaking population for 1990.

Need "Data Set 12: Number of Non–English-Speaking Americans"

How the United States Has Grown

What is the relationship between year and United States' population over the period 1790–1990?

Finding a Relationship

Organize

- Use "Data Set 13: Population of the United States from 1790–1990," page 142. Enter the dates as the *x* variable and the populations (in millions) as the *y* variable (see "Calculator Help 1: Entering Data," page 110).
- Test a linear relationship and various transformations to decide which best models the data.
- Determine one or more best-fit equations.

Communicate

- Describe the process you used to model the data.
- Can more than one model be justified? Explain.

Reflect

- Use your model(s) to predict the United States' population for the year 2000.
- Could a better prediction for the year 2000 be obtained by using only a subset of the data? Explain.

Need "Data Set 13: Population of the United States from 1790–1990"

Looking Back

The key concepts in this module are listed below. Using appropriate descriptions, illustrations, or examples, summarize two or more of these in such a way that you could explain them to others in your class.

Scatter Plot
Linear Relationship

Line of Best Fit
Correlation and Linear Correlation Coefficient
Error and Mean Square Error

Predicting and Forecasting
Nonlinear Data and Transformations of Data

Appendix A

Blackline Masters

Blackline Master 1: Circle Mats
(For Developing 2-6)

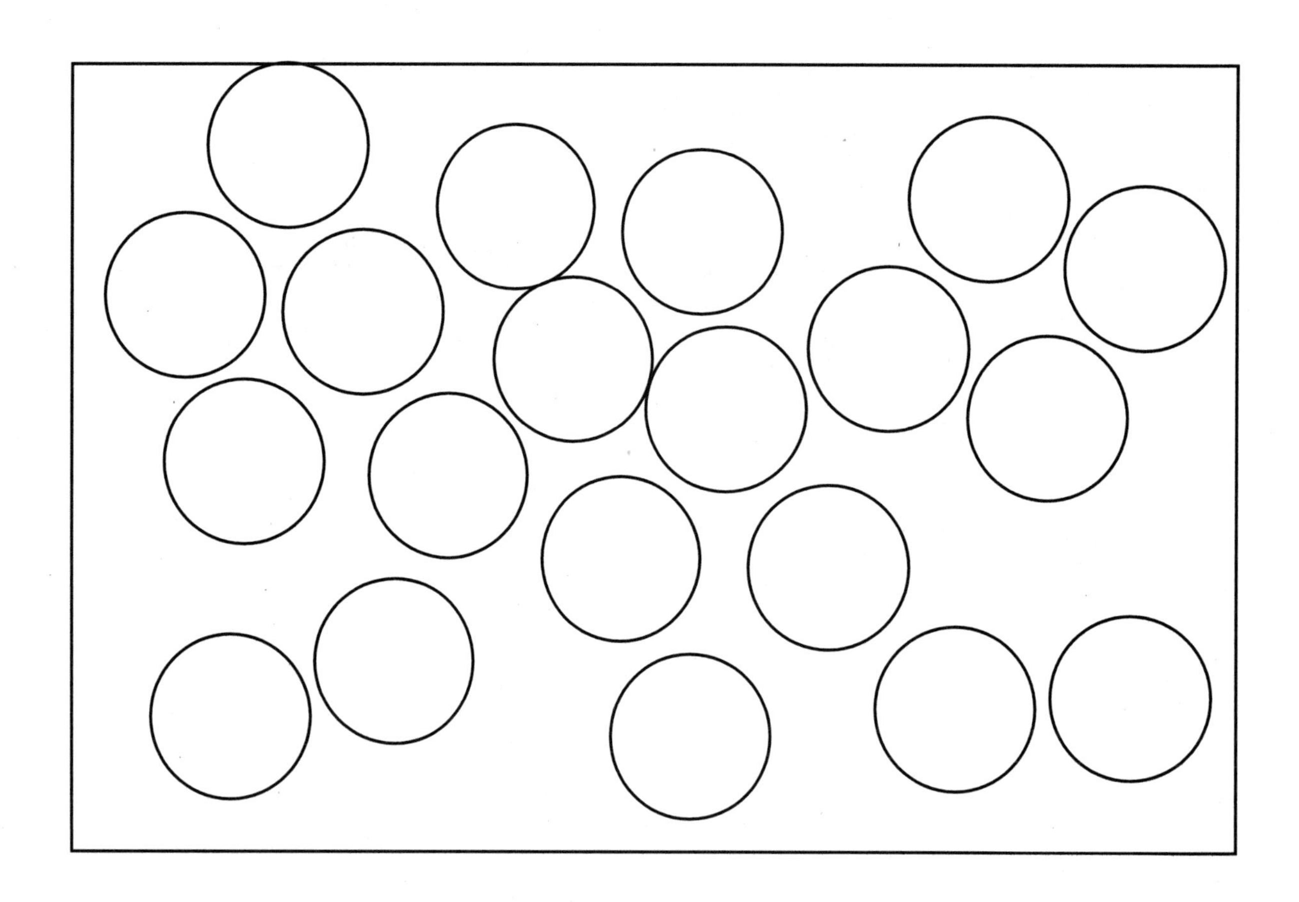

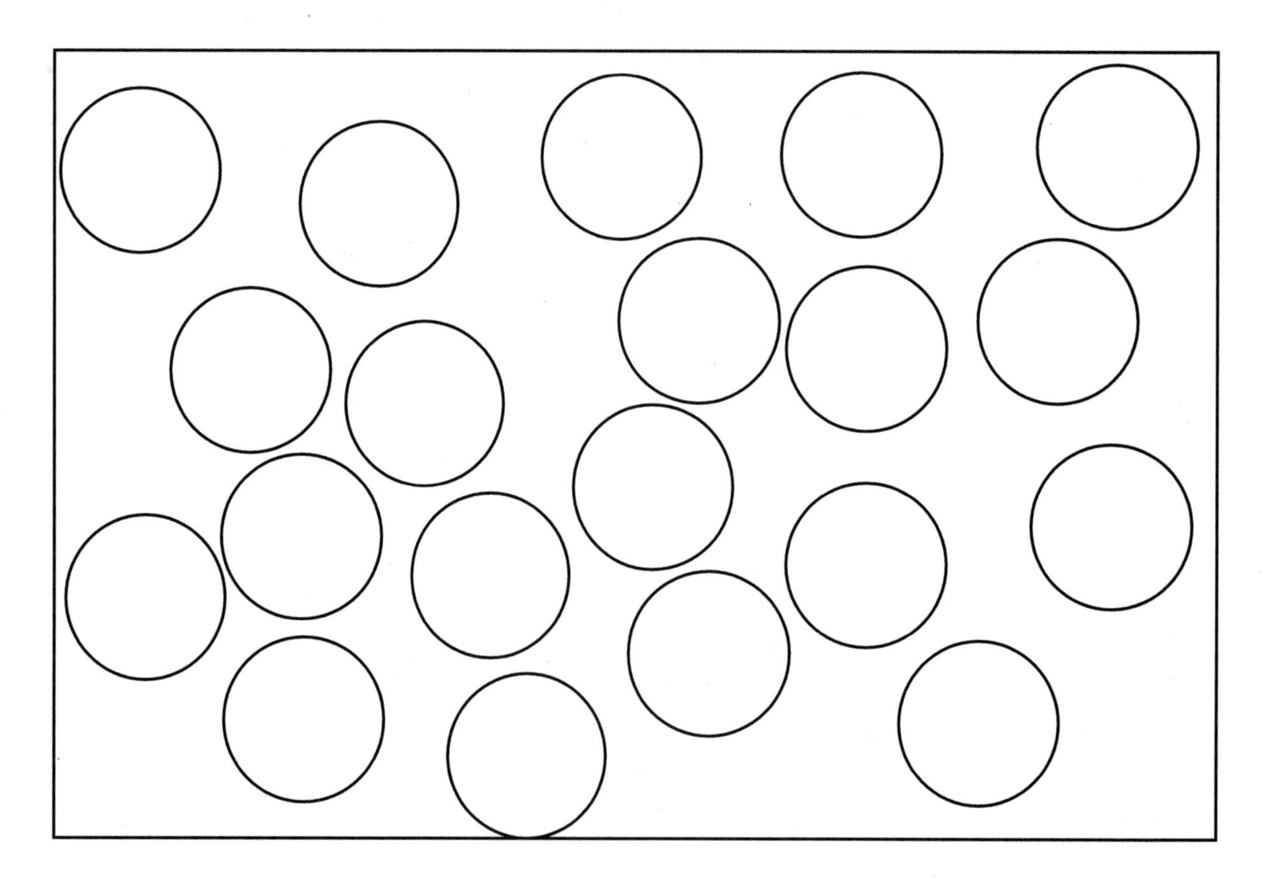

Kimberly (February)	Stacey (July)	Graham (March)	Timothy (May)
Cameron (November)	Kimiro (April)	Heather (August)	Jim (January)
Michiyo (June)	Jody (September)	Toni (October)	Allen (July)
Cedric (May)	Neil (January)	Keran (December)	Kristina (April)
Natalia (February)	Corchita (August)	Katarina (November)	Andre (September)
Shelly (December)	Ria (October)	Leigh (March)	Rudy (June)

Blackline Master 3: More Names and Birth Months
(For Starting 3-4)

Sheryl (December)	Stephan (February)	Ettore (October)	Jiovanni (July)
Abhijit (September)	Zena (May)	Tara (March)	Troy (January)
Sophronia (April)	Yolanda (January)	Adolf (November)	Birgit (August)
Hector (March)	Michelle (June)	Blake (October)	Deshona (December)
Dominick (July)	Atteo (January)	Thanh (April)	Anthony (August)
Ching (December)	Katya (September)	Ivan (February)	Elanie (March)
Eualdene (October)	Maria (April)	Pablo (June)	Erlecia (November)

Blackline Master 4: Quarter-Inch Graph Paper
(For Developing 4-1)

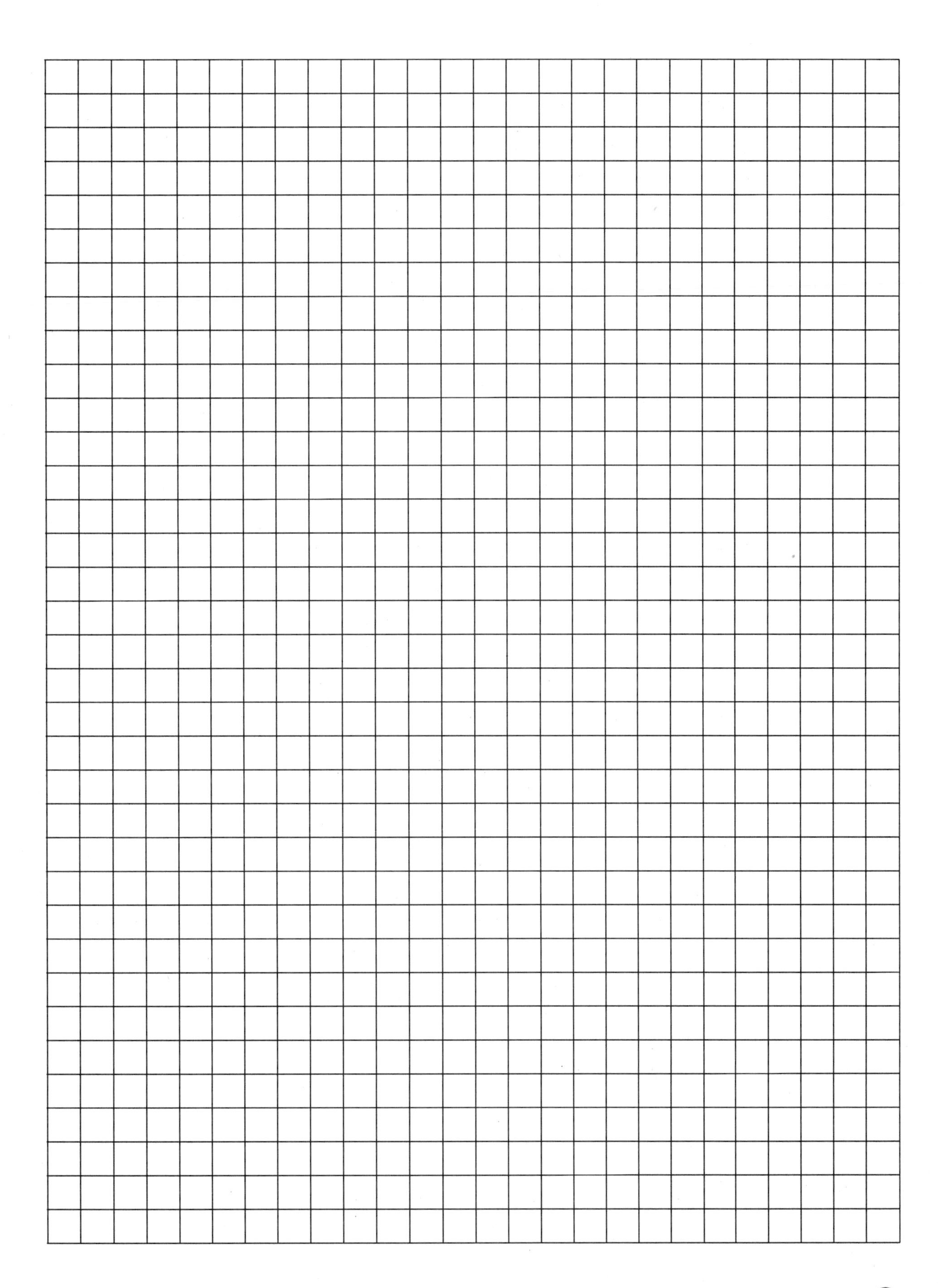

Blackline Master 5: Circles
(For Extending 4-2)

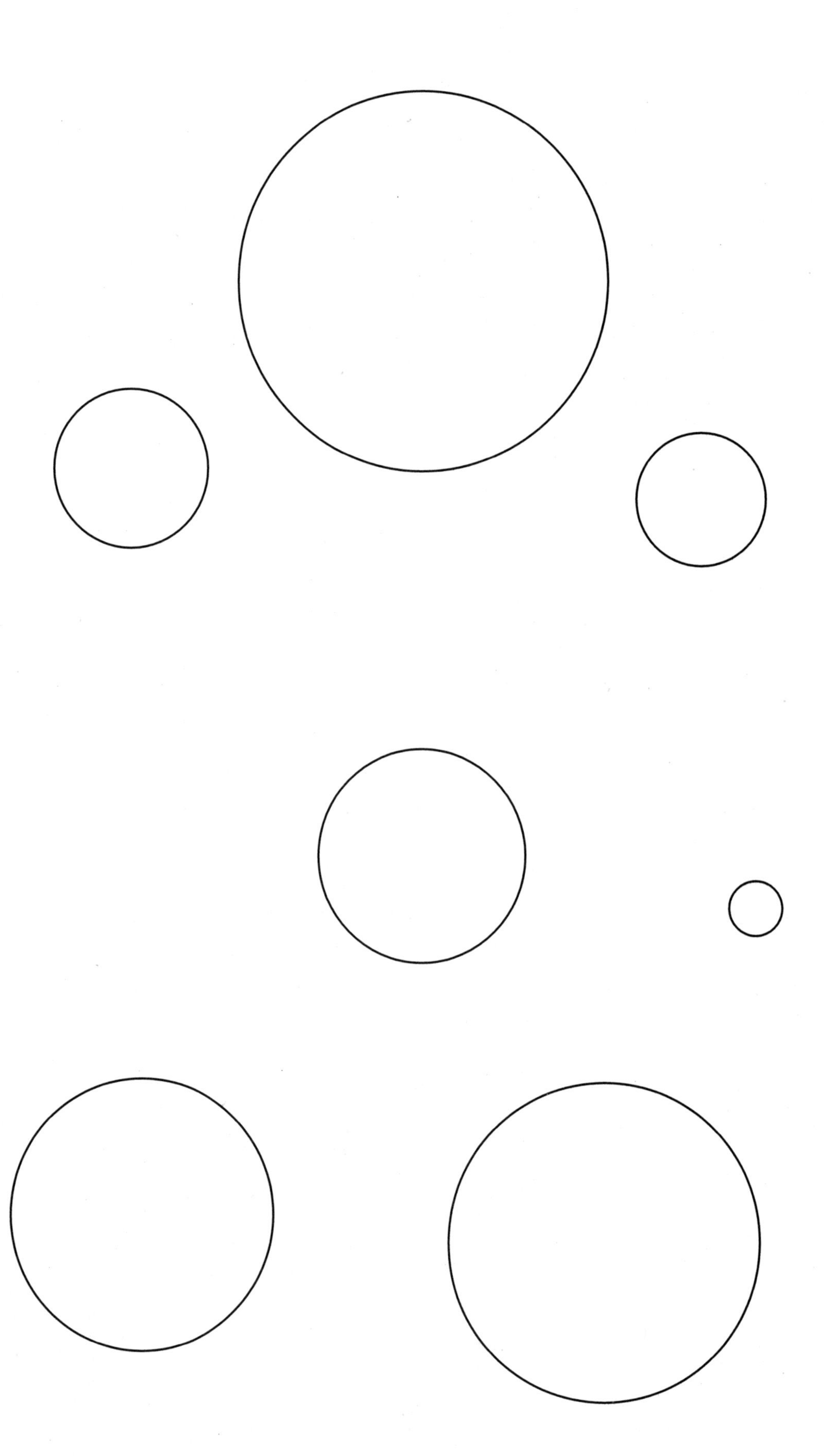

Appendix B

Calculator Helps

Single-Variable Data

- Clear any data in calculator by using the following sequence of key presses:

  1. Press [STAT] to display the STAT EDIT menu.
  2. Press [v] three times to select **4:ClrList** and press [ENTER] (or just press **4**).
  3. Press [2nd] [L1] [,] [2nd] [L2] [,] [2nd] [L3] [,] [2nd] [L4] [,] [2nd] [L5] [,] [2nd] [L6] to tell the calculator which lists to clear.
  4. Press [ENTER] and the message **Done** appears. All data have now been cleared from memory.

- To enter single-variable data:

  1. Press [STAT] to display STAT EDIT menu.
  2. Press [ENTER] to select **1:Edit.**
  3. The cursor will appear under L_1. Enter your values and press the [ENTER] key after each value. (If the frequency of your data is greater than one, the frequency values can be placed in the corresponding positions in L_2.) Continue until all the data are entered.
  4. The position of the entry in the list is shown at the bottom of the screen.

- When you are done entering your data in the list:

  1. Press [STAT] to get back to the statistics menu or [2nd] [QUIT] to quit the process.

Two-Variable Data

- To enter two-variable data:

  1. Follow the steps above entering the x values in L_1 and the corresponding y values (in the same position) in L_2.

Further details can be found in the TI-82 graphics calculator manual on pages 12-9 through 12-12.

Range

- The `WINDOW` key just under the screen gives the range values. The `WINDOW` determines the boundaries of your screen. The default values for Xmin, Xmax, Ymin, and Ymax are –10, 10, –10, and 10.
- To change any value in the `WINDOW`:

  1. Press the `WINDOW` key and then press `ENTER`. The cursor will be blinking on the <Xmin> value.
  2. Use the `v` or `^` arrow keys to move the cursor to the value that you want to change.
  3. Enter the new value. Your new value will delete the old value. If your value is negative, be sure to use the gray `)` key.
  4. Press `ENTER`. The cursor will move to the next value in the list.
  5. When you are finished, press `2nd` [QUIT] to quit the `WINDOW` menu.

Histogram

- Before drawing the histogram check that the following is done:

  1. Your data are entered.
  2. Your `WINDOW` values are appropriate for your data.
  3. The Xscl is one or greater as its value determines the width of the bars in the histogram.
  4. The `Y=` menu has no active functions. An active function has the = highlighted. If there are active functions, use the arrow keys to move the cursor on each active = and press `ENTER`. When you move the cursor the = will no longer be highlighted and the function will not graph. You may also place the cursor on the first character of the function and press `CLEAR`. This erases the function.

- To draw a histogram press the following key sequence:

  1. Press `2nd` [STAT PLOT] to display the STAT PLOT menu.
  2. Press `ENTER` to display the **Plot1** menu.
  3. Press `ENTER` again to turn the plot on.
  4. Press the down arrow `v` to highlight the first option by **Type:**.

5. Press the right arrow [>] three times to highlight the histogram and press [ENTER].
6. Press the down arrow [v] to highlight the first **Xlist:** option. If your data is in List1, press [ENTER]. Otherwise, use the arrow keys to highlight the list containing your data.
7. Press [v] to highlight the first **Freq:** option. If your frequency is **1**, press [ENTER]. Otherwise use the arrow keys to move to the list containing the frequency and press [ENTER].
8. Press [GRAPH] to show your histogram.

Further details can be found in the TI-82 graphics calculator manual on pages 3-8 to 3-9 and 12-19 to 12-21.

- The mean can be found in the 1-Var statistics in the STAT CALC menu. After the data have been entered, press the following keys:

1. Press [STAT] [>] to display the STAT CALC menu.
2. Press the down arrow twice to highlight **3:SetUp** and press [ENTER] (or press **3**).
3. The cursor is on L_1 in the Xlist. If the *x* variables are entered into L_1 press the down arrow [v] to go to frequency.
4. If the *x* values are in a different list use the right arrow [>] to highlight the list the press [ENTER] before you press the down arrow [v].
5. If the frequency is one, you are finished. If the frequency is in a list use the right arrow [>] to highlight the list number and press [ENTER].
6. The calculator screen will give the following information:
   1-Var Stats
   Xlist: L_1 L_2 L_3 L_4 L_5 L_6
   Freq:1 L_1 L_2 L_3 L_4 L_5 L_6

- You should be sure that the list containing your *x* values is highlighted and the frequency (one or a list number) is highlighted.
- To find the mean:

1. Press [STAT] [>] to display the STAT CALC menu.
2. Press [ENTER] [ENTER]. The following will appear on the screen:

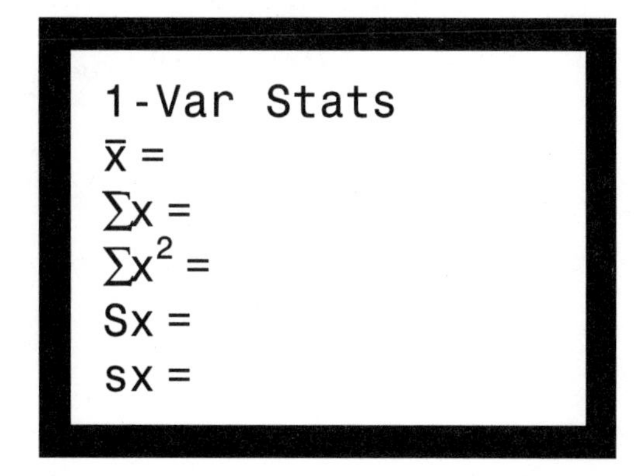

- The mean is the value of $\bar{x}$. You can check it by dividing Σx (the sum of the *x* values) by *n* (the number of *x* values).

Further details can be found in the TI-82 graphics calculator manual on page 12-13.

Calculator Help 4: Generating a Scatter Plot

- Before drawing the scatter plot make sure the following is done:

  1. Your data are entered into L_1 and L_2.
  2. The range in the WINDOW is appropriate for your data. You can use the following key sequence to get the appropriate WINDOW :

     a. Press ZOOM **9** (to select **9:ZoomStat**)

  3. The Y= menu has no active functions. An active function has the = highlighted. If there are active functions use the arrow keys to move the cursor on each active = and press ENTER . When you move the cursor, the = will no longer be highlighted and the function will not graph. You may also place the cursor on the first character of the function and press CLEAR . This erases the function.

- To draw your scatter plot press the following key sequence:

  1. Press 2nd [STAT PLOT] to display the STAT PLOT menu.
  2. Press ENTER to select **1:Plot1**.
  3. Press < ENTER to turn the plot on.
  4. Press v to move to the type and press ENTER to select the scatter plot.
  5. Press v and use the arrow keys to highlight the list containing your data and press ENTER .
  6. Press v and use the arrow keys to highlight the list containing your *y* data and press ENTER .
  7. Press v and highlight the mark you wish to appear on the graph.
  8. Press GRAPH to show your scatter plot.

Further details can be found in the TI-82 graphics calculator manual on pages 12-18 and 12-20.

- The random-number generator returns a number greater than 0 and less than 1, such as .9954663411 or .2003402618.
- To generate random numbers use the following sequence of key presses:

1. Press [MATH] to display MATH menu.
2. Press [<] to display the MATH PRB menu.
3. Press [ENTER] to select **1:Rand** (the random-number generator). **Rand** appears with the blinking cursor.
4. Press [ENTER] again and a random number appears. If you continue to press [ENTER], random numbers will continue to appear.

Further details can be found in the TI-82 graphics calculator manual on page 2-12.

Using the Line Function

- First draw your scatter plot (see "Calculator Help 4: Generating a Scatter Plot," page 114). Then press the following key sequence:

  1. Press [2nd] [DRAW] to display the DRAW menu.
  2. Use the down arrow [v] once to highlight **2:Line(**.
  3. Press [ENTER] to select **2:Line(**. A "+" will appear.
  4. Use the arrow keys to locate a starting position for the line you want to draw.
  5. Press [ENTER] to set the point at the first point on the line.
  6. Use the arrow keys to extend the line from the first point to the end position for the line.
  7. Press [ENTER] to set the endpoint. Your line will be drawn between the two points you selected using the "+."
  8. To cancel the Line(function press [CLEAR].

Cursor Movement in the Line Function

- Once [ENTER] has been pressed the second time in the line function (see above), the blinking cursor is positioned on the second endpoint chosen for your line. The x value and y value of that point is shown at the bottom of the screen.
- To find a point on the line:

  1. Use the arrow keys to move the cursor to the approximate x value you want.
  2. Use the arrow keys to move the cursor to the line.
  3. Read the approximate y value that is shown on the bottom right on the screen.

- To predict the arm length move the cursor until the x value is approximately 35. Then move the cursor to the line and read the y value.
- If you are given the y value:

  1. Use the arrow keys so the approximate value for y is at the bottom of the screen.
  2. Use the arrow keys to position the cursor on the line.
  3. Read the approximate x value from the bottom left of the screen.

Further details can be found in the TI-82 graphics calculator manual on page 8-4.

Finding the *y*-Intercept, the Slope of the Line of Best Fit, and the Correlation Coefficient (*r*)

- The *y*-intercept and slope of the line of best fit as well as the correlation coefficient can be found using the STAT CALC menu. To be sure the information is correct, you start with the <SetUp>:

  1. Press [STAT] [>] to display the STAT CALC menu.
  2. Press **3** to select **3:SetUp**.

- The line of best fit uses two-variable statistics. Use the <SetUp> to specify the two-variable statistics for your line. Use the arrow keys and [ENTER] key to highlight the correct values under **Xlist:**, **Ylist:**, and **Freq:**.

- After doing the setup, press the following to get the linear regression information:

  1. Press [STAT] [>] to display the STAT CALC menu.
  2. Press 5 to select **5:LinReg(ax + b)**.
  3. Press [ENTER] to display the <LinReg> values for *y*, *a*, *b*, and *r* where *b* is the *y*-intercept, *a* is the slope, and *r* is the correlation coefficient. These values and the equation are stored in the VARS menu.

Using the Equation to Draw the Line of Best Fit

- When you find the values of the slope and *y*-intercept of the line of best fit, the calculator stores those values as well as the equation of the line in memory as variables. The equation of the line of best fit can be found in the VARS menu.

- To draw the line of best fit do the following:

  1. Draw the scatter plot (see "Calculator Help 4: Generating a Scatter Plot," page 114).
  2. Press the [Y=] key. Press [CLEAR] if there is any formula already listed at y_1.
  3. With the cursor at y_1 press the following sequence of keys:
     a. Press [VARS] to display the VARS menu.
     b. Press **5** to select **5:Statistics ...**.
     c. Press [>] [>] to display the VARS EQ menu.
     d. Press **7** to select **7:RegEQ**.
     e. Press [ENTER] to copy the equation to the y_1 location.
     f. Press [GRAPH] to show the line on the scatter plot.

Further details can be found in the TI-82 graphics calculator manual on pages 12-3 to 12-8, 12-15, and 12-16.

- Once you have found the line of best fit (or line of regression), the calculator stores the equation in the VARS menu under EQ (see "Calculator Help 7: Finding the *y*-Intercept, the Slope of the Line, . . . Using the Equation to Draw the Line of Best Fit," page 117).
- To use the best-fit equation to predict a *y* value for a given *x* value, do the following:

1. From the text screen enter the *x* value that will be used for the prediction. Use the following key presses:

   a. Press the number keys to enter the number for the *x* value. For example, if the number is 35, press **3 5**.

   b. Press [STO▶] then [X] to store the value in *x*.

   c. Press [ENTER].

2. With the cursor on the text screen press the following sequence of keys:

   a. Press [VARS] **5** to display the VARS statistics menu.

   b. Press [>] [>] to display the VARS EQ menu.

   c. Press **7** to select **7:RegEQ** and to copy the equation to the text screen.

   d. Press [ENTER] again to evaluate the current value of *x* which was input.

Further details can be found in the TI-82 graphics calculator manual on pages 12-13, 12-15, and 12-16.

- Any data that are entered into a list can be easily transformed. This transformed data can be stored in another list.
- Enter the following sample data into L_1.

$L_1(1) = 6$	$L_1(6) = 19$
$L_1(2) = 8$	$L_1(7) = 25$
$L_1(3) = 10$	$L_1(8) = 29$
$L_1(4) = 12$	$L_1(9) = 35$
$L_1(5) = 15$	$L_1(10) = 51$

- If these data are transformed by using the logarithm function with the transformed data stored in List 3, you would press the following keys:

1. Press [STAT] [ENTER] to select **1:Edit** from the STAT EDIT menu.
2. Press [>] [>] to move to List 3.
3. Press [∧] to highlight L_3.
4. Press [LOG] [2nd] [L1] to display **L3=log L1** at the bottom of the screen.
5. Press [ENTER] to display the logarithm values of the corresponding data points from List 1 to List 3. List 3 should now contain the following values:

$L_3(1) = .77815$
$L_3(2) = .90309$
$L_3(3) = 1$
$L_3(4) = 1.0792$
$L_3(5) = 1.1761$
$L_3(6) = 1.2788$
$L_3(7) = 1.3979$
$L_3(8) = 1.4624$
$L_3(9) = 1.5441$
$L_3(10) = 1.7076$

- You can transform the values in any of the lists and place the transformed values in any other list by moving the cursor so it highlights the label of the desired storage list and following the above directions.

Further details can be found in the TI-82 graphics calculator manual on page 12-7.

Appendix C

Calculator Programs

Introduction

A program for the TI-82 graphics calculator is a set of instructions that the calculator can perform in order. The instructions are entered into the memory from the keyboard and menus of the calculator. Once the program is entered it can be run over and over because the calculator remembers the instructions.

The [PRGM] key accesses the three program menus. The first menu is EXEC. This is the menu from which you run a program. You use the arrow keys to highlight the program that you want to run and then press [ENTER]. The second menu is EDIT, and it allows you to correct a program in memory. The third menu is NEW. This is the menu that is used to enter programs into the calculator for the first time.

To enter a program press [PRGM] and then the right arrow twice. Press [ENTER] with **1:Create New** highlighted. The calculator will show **PROGRAM** with **Name=** under it. The blinking cursor is in "alpha" mode (you can type the gray letters shown above the keys) so you can type in a name for your program. It can be up to eight characters long. Once you have named a program you cannot change the name. When you press [ENTER] the cursor will move to the next line following a colon (**:**). The colon indicates the beginning of a new line of instructions for the calculator. You are now ready to begin entering your program. You enter each instruction individually and press [ENTER] when you have completed the instruction. An instruction may be longer than one line; just keep entering the information as it will wrap to the next line.

Most of the commands used are found in the menus that you have used. You just press the menu that contains the instruction that you want, use the arrows to highlight the instruction, and press [ENTER]. The calculator will place the appropriate command on the current line.

Pressing [PRGM] while you are in EDIT mode will give you three new menus. The CTL menu, the first of these, provides commands that help control the flow of the program. Such things as "label," "goto," and "end" are found in this menu. The second menu is I/O for input/output. This menu lets you display graphs and information and ask for input from the keyboard. The last menu is EXEC. You can select another program from this menu. When the calculator reaches a statement containing a program name, it will run the program listed.

Remember to begin each instruction on a new line and to press ENTER at the end. When you have finished entering the program, you can press 2nd [QUIT] to quit. The calculator will save your program for you. To run your program press PRGM and use the arrow keys to highlight the program you want to run from the EXEC menu. Press ENTER. The calculator will display the name with a blinking cursor. Press ENTER again and the program will run. The calculator looks for errors in your instructions when the program is run, not when you enter them. If there is a problem the calculator will display a message and you can go to the place in the program where the error occurred.

For example, the instructions for entering the program called TEN are given below.

`enter program name`	PRGM > > ENTER [T] [E] [N] ENTER
`:1→T`	**1** STO▶ ALPHA [T] ENTER
`:4→Q`	**4** STO▶ ALPHA [Q] ENTER
`:If I>N`	PRGM ENTER ALPHA [I] 2nd [TEST] v v ENTER ALPHA [N] ENTER
`:Goto A4`	PRGM **0** ALPHA [A] **4** ENTER
`:While I≤N`	PRGM **5** ALPHA [I] 2nd [TEST] **6** ALPHA [N] ENTER
`:int (L1(I)/10)→S`	MATH > **4** (2nd [L1] (ALPHA [I]) ÷ **1 0**) STO▶ ALPHA [S] ENTER
`:If S≥10`	PRGM ENTER ALPHA [S] 2nd [TEST] **4 1 0** ENTER
`:Then`	PRGM **2** ENTER
`:Goto A4`	PRGM **0** ALPHA [A] **4** ENTER
`:End`	PRGM **7** ENTER
`:If I≥2`	PRGM ENTER ALPHA [I] 2nd [TEST] **4 2** ENTER
`:Then`	PRGM **2** ENTER
`:int (L1(I−1)/10)→V`	MATH > **4** (2nd [L1] (ALPHA [I] − **1**) ÷ **1 0**) STO▶ ALPHA [U] ENTER
`:End`	PRGM **7** ENTER
`:If S>V`	PRGM ENTER ALPHA [S] 2nd [TEST] **3** ALPHA [V] ENTER

:Then	PRGM 2 ENTER
:W+1→W	ALPHA [W] + 1 STO▶ ALPHA [W] ENTER
:1→T	1 STO▶ ALPHA [T] ENTER
:4→Q	4 STO▶ ALPHA [Q] ENTER
:End	PRGM 7 ENTER
:If T=1	PRGM ENTER ALPHA [T] 2nd [TEST] ENTER 1 ENTER
:Then	PRGM 2 ENTER
:Output(W,2,5)	PRGM > 6 ALPHA [W] , 2 , 5) ENTER
:Output(W,3,":")	PRGM > 6 ALPHA [W] , 3 , ALPHA ["] 2nd [:] ALPHA ["]) ENTER
:T+1→T	ALPHA [T] + 1 STO▶ ALPHA [T] ENTER
:End	PRGM 7 ENTER
:Output(W,Q,L1(I)−10S)	PRGM > 6 ALPHA [W] , ALPHA [Q] , 2nd [L1] (ALPHA [I]) − 1 0 ALPHA [S]) ENTER
:I+1→I	ALPHA [I] + 1 STO▶ ALPHA [I] ENTER
:Q+1→Q	ALPHA [Q] + 1 STO▶ ALPHA [Q] ENTER
:If I>N	PRGM ENTER ALPHA [I] 2nd [TEST] 3 ALPHA [N] ENTER
:Then	STO▶ 2 ENTER
:Goto A4	PRGM 0 ALPHA [A] 4 ENTER
:End	PRGM 7 ENTER
:End	PRGM 7 ENTER
:If I+1 > N	PRGM ENTER ALPHA [I] + 1 2nd [TEST] 3 ALPHA [N] ENTER
:Then	PRGM 2 ENTER
:Goto A4	PRGM 0 ALPHA [A] 4 ENTER
:End	PRGM 7 ENTER
:Lbl A4	PRGM 9 ALPHA [A] 4 ENTER
:Return	PRGM ALPHA [E] ENTER

You may now press 2nd [QUIT] to quit. The program TEN will be saved. To run this program simply press PRGM, use the arrow keys if necessary to highlight the number associated with TEN, and press ENTER. The calculator will respond with program TEN with the cursor blinking at the end. When you press ENTER the program will run. Other programs, such as those on the following pages, can be entered and run in a similar manner.

The Stem-and-Leaf Program (page 125) takes a list of data points (L1) and sorts it, finds the maximum and minimum values, and uses that information to create a stem-and-leaf plot. The programs TEN and HUN (shown on pages 127 and 128) are sub-programs of the Stem-and-Leaf Program and must be entered first. TEN and HUN set up stem-and-leaf plots for data that is between 0 and 99, and data that is between 100 and 999, respectively.

Further instructions for programming on the TI-82 can be found in the TI-82 manual on pages 13-1 to 13-18. Sample programs appear on pages 14-8, 14-10, and 14-17. Also helpful for finding functions for your program are the "Table of Functions and Instructions" beginning on page A-2 and the menu map beginning on page A-22. Texas Instruments also publishes *Introduction to Programming on the TI-82.*

Calculator Program 1: Stem-and-Leaf Plot

name the program STEMLEAF

Program	Comment
:ClrHome	(**ClrHome** is under the PRGM I/O menu as number 8)
:SortA(L1)	Sorts List 1 (**SortA(** is under [2nd] [LIST] menu)
:L1(1)→M	*Stores minimum value*
:Input N	
:L1(N)→T	*Stores maximum value*
:ClrHome	*Clears screen*
:1→W	
:1→I	*Initializes values*
:1→T	
:4→Q	
:If M>9	*Checks size of minimum*
:Then	
:Goto A1	
:End	
:If I≤N	*Checks values*
:Then	
:While L1(I)≤9	*Starts loop to check values ≤ 9*
:If T=1	
:Then	
:Output(1,2,"0:")	*Outputs the stem 0*
:End	
:If I>N	
:Then	
:Goto A3	
:End	
:Output(1,Q,L1(I))	*Outputs values after stem*
:I+1→I	
:2→T	
:Q+1→Q	
:End	*Stops loop*
:End	
:4→Q	
:If I>N	
:Then	
:Goto A3	
:End	
:Lbl A1	
:If M>99	*Checks size If > 99 goes to new program*
:Then	
:Goto A2	
:End	
:PrgmTEN	*Calls program TEN*

```
:If I>N
:Goto A3
:Lbl A2
:PrgmHUN                    Calls program HUN
:Lbl A3
:Stop                       Ends program
```

The authors gratefully acknowledge the assistance of John Hill, Lincoln College, Normal, Illinois, with this program.

Calculator Program 2: Ten

```
Name the program TEN
:1→T                              Initializes values
:4→Q
:If I>N
:Goto A4
:While I≤N                        Sets loop to find stem and leaf
:int(L1(I)/10)→S                  Finds stem
:If S≥10
:Then
:Goto A4
:End
:If I≥2
:Then
:int (L1(I–1)/10)→V
:End
:If S>V                           Checks size
:Then
:W+1→W
:1→T
:4→Q
:End
:If T=1
:Then
:Output(W,2,S)                    Outputs values
:Output(W,3,":")
:T+1→T
:End
:Output(W,Q,L1(I)–10S)
:I+1→I
:Q+1→Q
:If I>N                           Checks size
:Then
:Goto A4
:End
:End
:If I+1>N                         Checks size
:Then
:Goto A4
:End
:Lbl A4
:Return
```

Calculator Program 3: Hundred

```
Name the program HUN
:1→T
:4→Q                      Initializes values
:If I>N
:Goto A5
:While I≤N                Starts loop to find stem and leaf
:int (L1(I)/10)→S         Finds values
:If S≥100                 Checks size
:Then
:Goto A5
:End
:If I≥2
:Then
:int (L1(I−1)/10)→V       Finds values
:End
:If S>V                   Checks size
:Then
:W+1→W
:1→T
:4→Q
:End
:If T=1
:Then
:Output(W,2,S)            Outputs values
:Output(W,3,":")
:T+1→T
:End
:Output(W, Q,L1(I)−10S)   Outputs values
:I+1→I
:Q+1→Q
:End
:If I>N                   Checks size
:Then
:Goto A5
:End
:End
:If I + 1>N               Checks size
:Then
:Goto A5
:End
:End
:Lbl A5
:Return                   Returns to STEM program
```

Appendix D

Data Sets

Data Set 1: First Names of Students

Kimberly
Stacy
Graham
Timothy
Cameron
Kimico
Heather
Jim
Michiyo
Jody
Toni
Allan
Cedric
Neil
Kevan
Kristina
Natalia
Conchita
Katarina
Andrew
Shelley
Ria
Sheryl
Leigh
Rudy

American League Statistics (1993)

Team	Games Won	Runs Scored
Baltimore	85	786
* Boston	80	686
* California	71	684
Chicago	94	776
Cleveland	76	790
Detroit	85	899
* Kansas City	84	675
Milwaukee	69	733
Minnesota	71	693
New York	88	821
Oakland	68	715
Seattle	82	734
Texas	86	835
Toronto	95	847

National League Statistics (1993)

Team	Games Won	Runs Scored
Atlanta	104	767
Chicago	84	738
Cincinnati	73	722
Colorado	67	758
Florida	64	581
Houston	85	716
Los Angeles	81	675
Montreal	94	732
New York	59	672
Philadelphia	97	877
Pittsburgh	75	707
San Diego	61	679
San Francisco	103	808
St. Louis	87	758

Source: *The 1994 Information Please Almanac.*

* When told to use a restricted list, omit those with an asterisk.

Data Set 3: First Names of Students and Their Shoe Lengths

	Number of Consonants	Shoe Length (cm)	Arm Length (cm)
Kimberly	6	23	31
Stacy	4	25	36
Graham	4	29	47
Timothy	5	31	50
Cameron	4	32	49
Kimico	3	21	30
Heather	4	24	32
Jim	2	28	44
Michiyo	4	22	31
Jody	3	27	40
Toni	2	29	46
Allan	4	26	38
Cedric	4	27	42
Neil	2	27	44
Kevan	3	29	47
Kristina	5	25	34
Natalia	3	21	29
Conchita	5	26	39
Katarina	4	21	28
Andrew	4	28	45
Shelley	5	24	33
Ria	1	23	32
Sheryl	5	26	38
Leigh	3	22	30
Rudy	3	28	46

Player		Money
1.	Steffi Graf	$2,821,337
2.	Arantxa Sanchez Vicario	1,938,239
3.	Conchita Martinez	1,208,795
4.	Martina Navratilova	1,036,119
5.	Gabriela Sabatini	957,680
6.	Jana Novotna	926,646
7.	Natalia Zvereva	857,160
8.	Gigi Fernandez	671,063
9.	Helena Sukova	655,573
10.	Mary Joe Fernandez	611,681
11.	Manuela Maleeva-Fragniere	561,320
12.	Amanda Coetzer	478,108
13.	Anke Huber	469,327
14.	Larisa Neiland	450,296
15.	Zina Garrison-Jackson	438,797
16.	Monica Seles	437,588

Source: *Tennis,* February 1994.

Data Set 5: The Ten Top-Selling Passenger Cars in the United States by Calendar Year (1989–1991)

Domestic and Import	Number Sold
1991	
1. Honda Accord	399,297
2. Ford Taurus	299,659
3. Toyota Camry	263,818
4. Chevrolet Cavalier	259,385
5. Ford Escort	247,864
6. Chevrolet Corsica/Beretta	231,227
7. Chevrolet Lumina	217,555
8. Honda Civic	205,715
9. Toyota Corolla	199,083
10. Ford Tempo	189,457
1990	
1. Honda Accord	417,179
2. Ford Taurus	313,274
3. Chevrolet Cavalier	295,123
4. Ford Escort	288,727
5. Toyota Camry	284,595
6. Chevrolet Corsica/Beretta	277,176
7. Toyota Corolla	228,211
8. Honda Civic	220,852
9. Chevrolet Lumina	218,288
10. Ford Tempo	215,290
1989	
1. Honda Accord	362,707
2. Ford Taurus	348,081
3. Ford Escort	333,535
4. Chevrolet Corsica/Beretta	328,006
5. Chevrolet Cavalier	295,715
6. Toyota Camry	257,466
7. Ford Tempo	228,426
8. Nissan Sentra	221,292
9. Pontiac Grand Am	202,185
10. Toyota Corolla	199,975

Source: *1993 World Almanac.*

Animal	mph	Animal	mph
Cheetah	70	Mule deer	35
Pronghorn antelope	61	Jackal	35
Wildebeest	50	Reindeer	32
Lion	50	Giraffe	32
Thomson's gazelle	50	White-tailed deer	30
Quarter horse	48	Wart hog	30
Elk	45	Grizzly bear	30
Cape hunting dog	45	Cat (domestic)	30
Coyote	43	Human	28
Gray fox	42	Elephant	25
Hyena	40	Black mamba snake	20
Zebra	40	Six-lined race runner	18
Mongolian wild ass	40	Wild turkey	15
Greyhound	39	Squirrel	12
Whippet	36	Pig (domestic)	11
Rabbit (domestic)	35	Chicken	9

Note: Most of these measurements are for maximum speeds over approximate quarter-mile distances. Exceptions are the lion and elephant, whose speeds were clocked in the act of charging; the whippet, which was timed over a 200-yard run (of 13.6 seconds); and the black mamba and six-lined race runner, which were measured over various small distances.

Source: *The World Almanac and Book of Facts,* 1994, p. 175.

Group	United States	Foreign
Mammals	37	249
Birds	57	153
Reptiles	8	64
Amphibians	6	8
Fishes	55	11
Snails	12	1
Clams	50	2
Crustaceans	10	1
Insects	13	4
Arachnids	3	1

Source: *The World Almanac and Book of Facts,* 1994, p. 175.

A group of 20 students in a local youth ensemble voted *Phantom of the Opera* and *Les Miserables* as their favorite Broadway shows. The students were polled to determine how many had actually seen the two shows. The results of the poll are recorded in the table below.

Name	Broadway Favorite*	
	Phantom	***Les Miserables***
Mikail	√	√
Mia	√	—
Jane	√	√
Jennifer	√	√
Ahkin	√	—
Jim	√	√
Orlyn	√	√
Rebecca	√	—
Ivan	√	√
Jana	—	√
Saad	—	—
Conchita	√	√
Pierre	—	√
Mercedes	√	√
Zach	√	√
Jamal	√	√
Anita	—	—
Orlie	√	√
Kimico	√	—
Lupe	√	√

* √ show was seen — show was not seen

In the first ten weeks of 1994, the top pop albums were:

1. *Music Box* by Mariah Carey
2. *Doggy Style* by Snoop Doggy Dog
3. *Music Box* by Mariah Carey
4. *Music Box* by Mariah Carey
5. *Music Box* by Mariah Carey
6. *Jar of Flies* by Alice in Chains
7. *Kickin' It Up* by John Michael Montgomery
8. *Toni Braxton* by Toni Braxton
9. *Music Box* by Mariah Carey
10. *Music Box* by Mariah Carey

Source: *Billboard* magazine, ten issues: January 8, 1994 through March 12, 1994.

Ranking	Movie	Money Made (billions)	Year Released
1.	*E.T. The Extra-Terrestrial* (Universal)	$ 228	1982
2.	*Star Wars* (20th Century-Fox)	193	1977
3.	*Return of the Jedi* (20th Century-Fox)	169	1983
4.	*Batman* (Warner Brothers)	150	1989
5.	*The Empire Strikes Back* (20th Century-Fox)	*	1980
6.	*Home Alone* (20th Century-Fox)	140	1990

* Has been omitted and will be predicted as part of an investigation.

Source: *The 1994 Information Please Almanac,* p. 750.

Data Set 11: Annual World Crude-Oil Production (1880–1992)

Year	Million Barrels
1880	30
1890	77
1900	149
1910	328
1920	683
1930	1412
1940	2150
1950	3803
1960	7674
1970	16,690
1980	21,722
1992	22,054

Sources: Data from the Energy Information Administration, recorded in Robert H. Romer, *Energy: An Introduction to Physics,* W. H. Freeman, San Francisco, 1976, for 1880 to 1970, *The World Almanac and Book of Facts 1991,* Newspaper Enterprise Association, New York, 1990, for 1980 and 1992.

Language	1980 Population 3 years and over	1990 Population 5 years and over
Spanish	11,549,333	*
French	1,572,275	1,702,176
German	1,606,743	1,547,099
Italian	1,633,279	1,308,648
Chinese	631,737	1,249,213
Polish	826,150	723,483
Vietnamese	203,268	507,069
Japanese	342,205	427,657
Arabic	225,597	355,150
Hindi and related	129,968	331,484
Russian	174,623	241,798
Yiddish	320,380	213,064
Navaho	123,169	148,530
Hungarian	180,083	147,902
Hebrew	99,166	144,292
Pennsylvania Dutch	68,202	83,525
Swedish	100,886	77,511
Cajun	9,374	33,670

* Has been omitted and will be predicted as part of an investigation.

Source: *The 1994 Information Please Almanac,* p. 830.

Data Set 13: Population of the United States from 1790–1990 (millions of persons)

Date	Population	Date	Population	Date	Population
1790	3.9	1860	31.4	1930	122.8
1800	5.3	1870	39.8	1940	131.7
1810	7.2	1880	50.2	1950	151.3
1820	9.6	1890	62.9	1960	179.3
1830	12.9	1900	76.0	1970	203.3
1840	17.1	1910	92.0	1980	226.5
1850	23.2	1920	105.7	1990	248.7

Source: United States Bureau of the Census as reported in *The World Almanac and Book of Facts,* 1990.

0. Do you think the graph of the data will be
   a) linea
   b) quad
   c) expon
   d) no correla
1. graph → show on calc
2. give equation
3. predict population in
   a) 2000 → 457.5
   b) 2010 → 563.13

Appendix E

Word Bank

back-to-back plot (Developing 1–4) A stem-and-leaf plot that displays two sets of data using the same stems. One set of data has the leaf displayed to the right of the stem, and the other set of data has the leaf displayed to the left of the stem.

box-and-whiskers plot (Developing 1–5) A graphical method of displaying data that is based on the quartiles of the data set. Quartiles divide the data set into four groups, each containing 25 percent of the measurements. The box is formed by the lower and upper quartiles with the median shown between these values. The whiskers are lines from the box that extend to the most extreme values.

combination (Developing 3–8) A combination, written ${}_nC_r$, equals the number of unordered arrangements of r objects taken from a set of n objects. ${}_nC_r = \frac{n!}{[(n-r)! * r!]}$. For example, ${}_{15}C_3 = \frac{15!}{[12! * 3!]} = 455$.

complement (Starting 2–9) The complement of an event E is the event that E does not happen. To find the probability of the complement, find $1 - \Pr(E)$. For example, let E be the event that we draw a five from a standard deck of cards. The complement of E is the event that we draw any card but a five. The probability that we do not draw a five (the complement of E) from a standard deck of cards is $1 - \frac{1}{13} = \frac{12}{13}$.

correlation (Developing 1–8) A measure of how two sets of values are associated. This association can be positive if both sets of numbers increase at the same time or negative if one set of data decreases at the same time as the other increases. If the numbers are not related, the correlation is said to be zero.

correlation coefficient (*r*) (Developing 4–5) A measure of how well the equation of the line fits the corresponding data. The value of r is a number between -1 and 1, inclusive. Values close to ± 1 indicate a good fit, while values close to 0 indicate a poor fit. Also known as the *linear correlation coefficient.*

dependent variable (Developing 1–7) The variable predicted when the independent variable is known.

equally likely outcomes (Starting 2–2) If the occurrence of each event in an experiment is as likely to occur as any other, then the events are called *equally likely outcomes.* If you toss a fair coin, a head is as likely to appear as a tail. So tossing a head and tossing a tail with a fair coin are equally likely outcomes.

experimental probability (Starting 2–1) A number that indicates the likelihood that an event will occur based on the results of an experiment.

factorial — (Developing 3–1) If n is a natural number the symbol $n!$ (read n factorial) equals the product of natural numbers less than or equal to n. For example, $5! = 5 * 4 * 3 * 2 * 1 = 120$.

frequency — (Starting 1–6) The number of times the data value occurs in the data set.

geometrical probability — (Developing 2–6) A probability based on geometric information and geometric measures of length, area, and volume.

histogram — (Starting 1–1) A bar graph that shows the relative frequency of the data for given intervals. The horizontal axis gives the intervals into which the data are divided and the vertical axis gives the proportion of the data that fall in each interval.

independent variable — (Developing 1–7) The variable used to determine the value of a second variable.

interquartile range (IQR) — (Starting 1–7) The difference between the upper and lower quartiles.

legend — (Developing 1–1) The legend of a stem-and-leaf plot shows the meaning of the stem and leaf for the particular plot.

line of best fit — (Starting 4–1) The line that most closely approximates a data set that is almost linear.

lower and upper quartiles — (Starting 1–5) The lower quartile is the 25th percentile of the data set. It represents the median of the lower half of the measurements. The upper quartile is the 75th percentile and represents the median of the upper half of the measurements in the data set.

mean — (Starting 1–3) The mean of a set of data is equal to the sum of the measurements divided by the number of measurements contained in the data set.

mean square error — (Developing 4–2) The average of the squares of the error values found from a line that models the data. (The error is found by taking the difference between each y value of the data points and the y value predicted by the line for that point, that is, the error is $(y - y')$. Mean square error $= \frac{[\text{sum } (y - y_i')^2]}{n}$ (where n is the number of points).

median — (Starting 1–4) The middle number when the measurements in the data set are arranged in ascending (or descending) order. When the data set has an even number of values, the median is the mean of the two "middle" values.

mode (Starting 1–6) The measurement that occurs most frequently in the data set.

odds (Developing 2–4) If the "odds in favor of an event A" are m to n, then the event A will occur m times for every n times it did not occur. The probability that A will happen is $\frac{m}{(m + n)}$.

outlier (Starting 1–2) An extreme value that represents relatively rare occurrences. Outliers lie at distances greater than (1.5 * interquartile range) beyond the upper and lower quartiles.

permutation (Developing 3–4) A permutation, written ${}_nP_r$, equals the number of *ordered* arrangements of r objects taken from a set of n objects. ${}_nP_r = \frac{n!}{(n-r)!}$. For example, ${}_{20}P_3 = \frac{20!}{17!} = 6840$.

random number (Starting 2–2) Number in which each of the digits zero through nine have an equal probability of being selected for any position in the number. Tables of random numbers are used to simulate experiments.

range (Starting 1–7) The difference between the largest measurement and the smallest measurement in the data set.

rounding (Developing 1–3) A number is rounded to a given place value when the digit in the designated position is determined using the following rules:

1) if the next digit to the right is zero through four, the designated value is unchanged;
2) if the next digit to the right is five through nine, the designated value is increased by one; and
3) all digits to the right of the designated position become zeroes.

For example, 358 rounded to the tens place would be 360, while 892 rounded to the tens place would be 890.

scatter plot (Developing 1–7) A graphical display of data that consists of two different values. The points representing pairs of data are plotted on a coordinate system.

simulation (Starting 2–2) The use of random numbers, spinners, or some other device to model an experiment or a real-world activity.

stem-and-leaf plot (Developing 1–1) A graphical method of summarizing a data set. Each measurement in the data set is divided into two pieces—the stem, which consists of digits to the left of a designated position, and the leaf, which consists of the remaining digits to the right of the stem.

theoretical probabilities (Starting 2–2) The probability values assigned to events based on mathematical theory rather than on trials of an experiment.

transformation (Extending 4–3) A change of an equation to an equivalent equation.

trial (Starting 1–1) One repetition of an experiment.

truncate (Developing 1–3) A number is truncated when all the digits following a given place value are deleted. For example, 345 truncated to the tens place would be 34.

upper and lower extremes (Starting 1–7) The upper extreme is the largest measurement in the data set while the lower extreme is the smallest measurement in the data set.

with replacement/without replacement In an experiment in which objects are drawn from a set, it is *with replacement* if the object that is drawn is returned to the set before another object is drawn. It is *without replacement* if the object drawn is not returned before the next draw.

References

Curriculum and Evaluation Standards for School Mathematics: Connecting Mathematics, Addenda Series Grades 9–12. Reston, Va.: National Council of Teachers of Mathematics, 1991.

Curriculum and Evaluation Standards for School Mathematics: Data Analysis and Statistics, Addenda Series Grades 9–12. Reston, Va.: National Council of Teachers of Mathematics, 1992.

Organizing Data and Dealing with Uncertainty, rev. ed. Reston, Va.: National Council of Teachers of Mathematics, 1979.

Dossey, John, et al. *Discrete Mathematics.* Glenview, Ill.: Scott, Foresman and Company, 1987.

Gnanadesikan, Mrudulla, Richard L. Scheaffer, and Jim Swift. *The Art and Techniques of Simulation.* Palo Alto, Calif.: Dale Seymour Publications, 1987.

Kenney, Margaret J., ed. *Discrete Mathematics Across the Curriculum, K–12: 1991 Yearbook.* Reston, Va.: National Council of Teachers of Mathematics, 1991.

Landwehr, James M., and Ann E. Watkins. *Exploring Data: Revised Edition.* Palo Alto, Calif.: Dale Seymour Publications, 1995.

Moore, David S., and George P. McCabe. *Introduction to the Practice of Statistics.* New York: W. H. Freeman and Company, 1993.

Newman, Claire M., Thomas E. Obremski, and Richard L. Scheaffer. *Exploring Probability.* Palo Alto, Calif.: Dale Seymour Publications, 1987.

Witmer, Jeffrey A. *Data Analysis: An Introduction.* Englewood Cliffs, NJ: Prentice-Hall, 1992.